THE COMPLETE CB RADIO

Richard Nichols is currently editor of *Custom Car*, Britain's most popular motoring magazine which has played a major part in bringing CB radio to the attention of the public. In autumn 1980, also under his editorship, Britain's first CB magazine, *Breaker*, was launched. Richard Nichols has actively campaigned for the legalization of CB on national and local radio, on television and in the press. He is a member of the CBA (Citizen's Band Association), the UKCBC (United Kingdom Citizen's Band Committee) and REACT (Radio Emergency Associated Citizen's Teams).

THE COMPLETE CB RADIO

Richard Nichols

Illustrated by Ian Sowerby

Star

A STAR BOOK
published by
the Paperback Division of
W. H. ALLEN & Co. Ltd

A Star Book
Published in 1981
by the Paperback Division of
W. H. Allen & Co. Ltd.
A Howard and Wyndham Company
44 Hill Street, London W1X 8LB

Copyright © 1981 Tall Stories

Performance specifications MPT 1320 and MPT 1321
© Crown copyright 1981
Reprinted with permission

Printed in Great Britain by
Anchor Press Ltd, Tiptree, Essex

ISBN 0 352 31014 6

CONTENTS

INTRODUCTION

Progress. Now there's a word to conjure with. Like so many other things – traffic jams, Party Political Broadcasts and Income Tax – it's a reasonably modern invention which is sometimes rather less appealing than its victims would like it to be. After all, it's progress which is responsible for nuclear bombs, so perhaps we haven't got an awful lot to thank it for.

But like it or not it seems to be here to stay and it even appears to be getting quicker all the time. Romans and similar were whizzing about in chariots a thousand years before anyone thought of sacking horses and inventing the motor car, but it was a scant 65 years after the Wright brothers lifted tentatively off the sands of Kittyhawk (in a flight which covered less distance than the wingspan of a jumbo jet) to the moment when Neil Armstrong took his giant leap on behalf of mankind, making Jules Verne redundant and the producers of *Star Wars* very rich indeed.

There's no doubt that things happen very much quicker than they used to. Always assuming that we are dealing only with the mechanics of Progress, that is. The trouble starts when Progress becomes the property of politicians, who are specially trained to take longer doing anything than the rest of us. It's hard to avoid the suspicion that if Progress had been left solely to governments we'd still be painting ourselves blue and living in hypocausts. Ask any archaeologist; not even the Romans could master the art of building above ground level, although some parts of Hadrian's Wall do appear to be an exception to this rule.

Consequently it has come as a considerable surprise to discover that the matter of CB Radio, for which we have, as they say, had the technology for some time, has passed through the hands of politicians in remarkably short order. It was only at the beginning of 1981 that the Government announced that it was at last not only in favour of, but committed to, the introduction of a CB service to this country and here we are now actually getting to grips with it all.

To be excruciatingly truthful there was a more than passing

amount of stupidity which preceded the acceptance of the principle by Authority, but even that only occupied three years or so which is, by Parliamentary standards, a mere pause for thought. After all, it took one lot six years to make up their minds about Hitler.

But the aggravation and the histrionics are all over now, and though they may have their place in legend the real story has yet to be written. It's a fact that the major ammunition fired by the pro-CB lobby during the campaign for legalization centred around the benefits to society which would accrue from the existence of a workable CB facility. Now it's here, society may at last settle down to enjoy them and put to practical usage all the advantages over which so many have waxed so eloquent for so long. For the housebound, the elderly, the injured, the lost and the lonely the CB revolution has only just begun.

1 THE HISTORY

It's all too easy to blame it on the Americans, whatever it is. Many years ago it was necessary to eviscerate chickens in order to ascertain the cause of evil and ensure that the correct person or persons met a horrible and grisly retribution. These days the blame for anything from Coca-Cola to Neutron Bombs may correctly be apportioned to America. Fortunately for Americans and the rest of the world responsibility for some things for which we may all be truly grateful is also theirs; for example the motion picture industry in general and some parts of it – Raquel Welch and similar – in particular. Also CB radio.

After the distressing incidents of the late eighteenth century it has always been difficult for anybody British to acknowledge the fact that in a lot of cases Americans are a great deal quicker on the uptake than we are. This doesn't extend to all matters of course – there's not a decent fast bowler among them – but it is especially true in the area of technology. As a rule we're pretty good at inventing stuff but we seldom seem to know what to do with it once we've got it, and in a great deal of cases we do tend to actually resist advancement rather than incorporate it.

In the light of that knowledge it should come as no surprise to find that the Federal Government legislated for the introduction of CB as long ago as 1948. At the time it wasn't called CB. It was the General Mobile Radio Service (GMRS) and it operated on the VHF frequency of 467MHz. The principle, however, was firmly established, and GMRS existed simply to provide access to a radio facility to the general public for whatever peaceful and law-abiding purpose they may have felt desirable at the time.

In itself this wasn't altogether a brand-new step, because the Amateur Radio (or Ham) network had existed on an international basis for some time. However, the terms of Ham licences are reasonably stringent, and the skill necessary to install and operate the equipment was (and still is) considerable. GMRS was designed to do away with all that. Morse code was unnecessary and in view of its designed-in short range all the internationally-recognized verbal shorthand was also redundant. All that was required was a

11

minimum quota of common sense and the ability to overcome the mike-fright which still afflicts beginners.

GMRS was not an outstanding success in terms of usage, however. The prime reason was that the liberal thinking which had provided the service was somewhat in advance of the technology of the day. The early sets were all valve-operated devices, noticeably larger than the transistorized stuff we're used to today and exceedingly sensitive. Although the world had proceeded beyond the cat's whisker radio sets the age of the spin-off hadn't yet properly arrived and GMRS in no way resembled the CB we have come to know and love.

It was 1958 before the major step forward took place and that owed its success to a small, insignificant-looking device; the transistor. This diminutive object had made its debut in 1949 but even then it was not the robust, hardy piece of equipment it has since become, and it was some time before commercially produced transistors could successfully be accommodated within electrical equipment designed for the rough treatment of the retail market. By the late fifties the old 'vibrator' valve-operated car radios were beginning to give way to the transistorized variety which were, by then, able to withstand the pounding which being fitted in a motor car metes out.

Even this was little help to GMRS, however. It's a fact that radio signals are easier to transmit at lower frequencies than high, and since GMRS was in the VHF (Very High Frequency) band you can see straight away that it was in big trouble right from the off. In order to fulfil the objectives of a public radio service the Americans introduced, in 1958, a second facility, Class D Citizen's Band Radio. This operated much further down the scale on 27MHz, which was by anybody's standards a rubbish frequency, used for all kinds of industrial equipment which pumps out a wide assortment of radio garbage at all times of the day or night. However, it was then much easier to utilize the frequency for voice transmissions than VHF, and American industry was able to meet the standards for these 23-channel radio sets quite adequately, for home base and mobile units alike.

Even this was ignored to begin with, and in view of the unreliability of the sets, coupled with the surprisingly varied range of electronic sound-effects of which they were capable, this will not, perhaps, come as much of a surprise to anyone.

12

But while the great percentage of the American public turned Nelsonian eyes towards CB, and saw no rigs, Progress was most definitely not in its hammock and a thousand miles away, although its attention was temporarily focused elsewhere. It was busy inventing the pocket calculator (presumably in order to facilitate the later introduction of VAT) via an unaccountably tortuous process which involved the microchip and the Apollo programme. The need to produce electronic equipment to the high standards necessary for the space race inevitably meant that much existing technology was swept away on an irresistible tide of advancement which left CB much better off than it had ever been before, and at last in a position to realize the potential for which it had been created.

You would be wrong if you thought that the late sixties ushered in a tremendous boom for American CB, although the foundations of the CB fraternity were well and truly laid by then. Topping out had to wait for a while though. To begin with it was the truckers who made the most use of the service, and who can blame them? To people like us, living on a tiny island, the concept of distance is entirely relative, and as far as long-haul trucking is concerned we are definitely at the kid brother end of the market. London to Glasgow may seem like a long way, but an American trucker may undertake journeys of hundreds of miles occupying several days, during which time he has nobody to talk to other than himself except at brief intervals.

The advent of CB changed all that, and allowed him contact with all kinds of people along his route; out on the long Interstate highways his companions on the airwaves were almost certain to be fellow-truckers. Doubtless you will not therefore be amazed to learn that the truckers established their own channel on the CB and built up a considerable array of jargon and slang which served several purposes; in the first place it must have made them laugh – you've only got to look through it to realize that – and it also identified them as part of a group, or club. To a lonely driver that last is probably more important than you'd think.

CB also provided a number of practical advantages as well. Aside from the time-saving factor allowed by the ease with which directions, either on a highway or within a strange town, may be obtained from someone on the air without the need to stop and ask, never mind get lost (you try taking a wrong turn on a motorway and see how much time it wastes), there were also benefits available following a breakdown. Sometimes it can be a long walk to the nearest phone in Surrey, never mind the Mojave Desert. And when time is money, as it is to the huge number of American truckers who are self-employed owner-drivers, things like that are more important than they are to the family motorist suffering the horrors and inconvenience of a breakdown between home and the seaside. And how many motorists have you seen go beserk when they take a wrong turning and have to spend two minutes turning round? Try doing a U-turn in any major city while you're driving a sixty-foot artic and trailer. CB was a very practical tool all right, as well as being a minor cult.

But boom time for the Americans didn't arrive until the OPEC countries discovered they had a crippling weapon coming out of holes in the ground, and all they had to do was put a plug in the end of the pipe. The fuel crises of the early seventies had many far-reaching effects and altered world attitudes to many things. Perhaps not the least significant of these was CB.

Hardly anyone will have failed to observe that American cars are built on an entirely different scale to the European variety, and even their 'compacts', which they are pleased to regard as small cars, are of impressive dimensions not totally dissimilar to the average aircraft carrier. The engines for these vehicles are constructed to a similar philosophy and consume gas, which is how Americans mistakenly describe petrol, in staggering quantities. The sudden disappearance of this commodity induced a state of near-hysteria among American motorists and there were even recorded instances in which they happily killed for it. The truckers had a more serious problem with fuel but they fortunately had a different solution to it which didn't require actual killing. Fuel to the trucker was more than his freedom and his leisure; it was how he earned his living. Locating supplies was therefore vital to his continuing commercial success and it was a short step to using CB to locate the precious stuff and inform other truckers of its whereabouts. Private motorists twigged this very swiftly and they all rushed out to buy CB rigs for their cars so they could also find petrol, and the biggest consumer boom to sweep America since the advent of colour TV was off and running.

Although the fuel shortages were only really bad for a while, supplies of petrol have never been the same in the States (some parts still operate the odd/even rationing system even now) but the use of CB as a sort of remote-control electronic divining rod is no longer as critical as it was. But strangely it didn't fade away as fuel supplies returned to what is now accepted as being normal. For one thing the Federal Government had decided that fuel was too precious to waste and imposed the blanket 55mph speed limit during the shortage and, by some oversight, neglected to cancel it afterwards. Speed limits, as we all know, are one of those 'occasional obedience' laws which we observe when it is convenient to do so or in the presence of a police car but frequently overlook when we are late home and know the only way of getting a hot dinner is by putting your foot down. Once again it was the truckers in particular who were made to suffer from this; the big rigs can cruise at far higher speeds than 55mph but can often take a long time getting there. And time is money when you're driving for a living. Consequently warnings about the presence of police cars or speed traps became commonplace on the truckers channel and everyone benefited. Even the police. The object of traffic police is prevention rather than maximum arrests, and their physical presence always slows down traffic to within the limit. If a warning went out on the air vehicles for miles in every direction slowed down, magnifying the effect tremendously. In fact it was such a good thing that many police forces still broadcast their own warnings on the CB, usually about fictitious speed traps. Saves manpower like crazy.

But apart from the practical benefits the motorists discovered that CB was a nice thing to have anyway, so they all got it. Millions of them. Demand for CB rigs multiplied almost weekly. By the mid seventies usage of the service was so great that the existing 23 channels were insufficient and the allocation was increased to 40. At the same time the technical specifications for the equipment were uprated somewhat to cope with the growing spectre of TVI – Television Interference – which had been gradually marring the honeymoon.

The old-standard sets were not outlawed, but they had to be off the dealers' shelves by January 1977. Towards the end of 1976 the supercheap bargain sale really got going and the old sets were practically given away free with boxes of matches before the hard sell on the new 40-channel equipment started. The strange thing about CB is that it is very nearly addictive and many of the millions who bought the cheap sets soon found that they would have been better off with the new ones. Being Americans, and therefore immensely wealthy, they went out and bought them in millions as well. In fact in 1977 they bought about ten million units, nearly all of which were being turned out by the Japanese, whose small nimble fingers are well-suited to the construction of miniature electronic devices at immense speed.

1977 was, as it turned out, the high point of the boom, and demand for sets fell off after that, finally settling at a steady two million units per year. Most of these are very probably replacement sets, as the rapid march of Progress (remember we're past the moon and halfway to Saturn or somewhere now) ushered in still more sophistication and miniaturization, but there are an estimated 40 million CB sets of one sort or another in the hands of the American public right now.

But while they're all busy using the things, the Japanese chappies have all been sitting around twiddling their nimble little fingers in frustration, because it's very tempting to create a production facility which can make ten million CB units every year but damn difficult knowing what to do with it afterwards. Also a problem knowing what to do with all the sets surplus to requirements sitting around in various warehouses all over the place. The answer, of course, is to expand your market somewhat, and if America doesn't want any more sets then someone else must. And if they don't then they should be made to. It's a simple theory of

salesmanship, but it's never been necessary to put it to use, since so many countries have seized on the CB subculture all by themselves, often despite the fact that it was illegal.

Most governments have been quick to spot the potential of CB and have legalized it in some form or another fairly swiftly, in nearly all cases using some form of specification based to greater or lesser degrees on the American system. It must be said that Britain was not in the forefront of this development. At the end of 1980 there were more than 60 countries in the world using CB on a legal basis, including most of Europe and one or two countries behind the Iron Curtain. We weren't one of them. And that wasn't through ignorance, or lack of demand. It was despite a full and detailed knowledge of the advantages the facility could bring and despite an intensive and long-running campaign aimed at the

17

introduction of legal CB as quickly as possible. Also despite radio piracy on an enormous scale.

Amateur Radio has long been an accepted part of radio usage in this country so it is safe to assume that governments are not totally opposed to the principle of allowing private citizens to use the airwaves, and in the light of that knowledge it is intriguing in the extreme to find their attitude towards CB has been so totally negative for so long. It's even more fascinating to discover that in the final event there was no special reason for it (or at least none that they would admit to) and that its introduction could have been made with little or no effort at any time within the past 30 years.

If you want a licence for anything other than a gun you simply trot along to your local Post Office and give them the money. They in turn give you a bit of paper which makes your TV, car, Ham radio or dog legally acceptable in the eyes of Authority. Dead simple. In all these cases the Post Office merely issues the licences on behalf of the appropriate Government department; it does only what it's told. As far as radio is concerned the Post Office is powerless to grant or deny permission for use. The right to do this is vested in Her Majesty's Minister of State for Home Affairs – the Home Secretary – and decisions about who to grant permission to, and for what purpose, are made by him after due consultations with the experts in the bit of the Civil Service concerned with the matter in hand.

Civil Servants, unlike politicians, are not accountable to the public. They are not required to seek a mandate from the electorate once every five years, or less if things are going badly. Consequently they tend to hold their jobs longer than the Ministers they serve. In addition they may find themselves advising both Conservative and Labour Ministers within a short space of time, and unless the subject in hand is one of party politics their advice, which is obviously going to be the same in both cases, will almost certainly be heeded by the Minister. This accounts for the similarity in attitude to some subjects held by Ministers of different political persuasion. More relevantly it explains why governmental attitude to CB has been the same for so long, despite the swings and roundabouts of political fortune and misfortune.

It is also true that any Minister will give his attention only to those things which he holds dear or which are so markedly in the public eye that they cannot be ignored. It is this question of priority which has led to the retention on the statute books of so many delightful but totally irrelevant laws. For example it's a fact that the playing of shove-groalie in Public Houses on Sundays is illegal. It's a law which Henry VIII passed and in which nobody has subsequently been sufficiently interested to either cancel or enforce.

The non-existence of CB for so long can then to a large extent be ascribed simply to disinterest rather than to opposition. In the main the rules regarding the use of radio transmitters were laid down in the Wireless Telegraphy Act of 1949, which principally says that nobody can transmit anything at all, with one or two detailed exceptions, and for a long time nobody particularly wanted it changed.

It wasn't until the mid-sixties that anybody in this country began to have even the faintest conception of what CB radio was, and even then it was hardly what you'd call a sudden awakening. The first CB sets to appear over here were introduced as toys – walkie-talkies, they were called – and were sold to parents of small boys, along with plastic Marine Corps hats, plastic Browning Automatic Rifles and instruction booklets teaching you how to impersonate John Wayne. These 100-milliwatt, single-channel handhelds require no licence at all in the States and have a maximum range only slightly better than a pair of strong lungs, but they were radio sets. Somewhere along the line someone saw a threat in these things and moves to have them banned were almost immediately successful. It would not, perhaps, be totally unfair to believe that British manufacturers of radio sets for industry were simply doing their best to protect their market before a flood of inscrutably cheap two-way radios arrived from the East. In any case 1968 saw the arrival of Statutory Instrument No. 61, 1968, Radiotelephonic Transceivers (Control of Importation and Manufacture). This prohibited absolutely the importation or manufacture of radio equipment operating on 27MHz. So even before any of us knew what it was the hand of Authority had deemed that CB was a bad thing and we mustn't have it.

This information came as a bit of a shock to the Charlie Bravo group, who'd known all about CB since 1965 and were having a fine time chatting to each other on their private radio network. They'd begun life on Channel 11, but interference had forced them up to Channel 14, where they stayed quite happily until the

19

early seventies, when a sharp spate of prosecutions forced them off the air.

They were soon replaced by another group – Lima Echo – who had moved up to the 23-channel mobile rigs available from the USA. Force of habit dictated that they would use Channel 14 as their net channel, and as their ranks gradually swelled 14 became universally accepted as the London calling channel, although most parts of the country adopted 19, the American truckers channel. And they began to adopt it in droves. Lima Echo suddenly lost itself in the welter of people – mostly young motorists – who had discovered the joys of CB, and who were enthusiastically rushing to join in.

And as they rushed to join in so Authority began rushing to stop them. In 1978 it was thought that there may have been as many as 500 illicit users of the 27MHz band, almost all of them in London. In the main these pirates used their sets late at night, generally sticking to the high ground in places like Epping, Richmond or Alexandra Palace. It was the latter which attracted the most attention, principally because there were so many radio transmitters or repeaters on top of the building, nearly all of which suffered some form of interference from the pirates.

To begin with the police visited these late-night breakers and asked them politely not to transmit and please to go away. This gentlemanly state of affairs lasted only a short while, however, principally because of the astoundingly rapid growth in the number of pirates. From 500 in late '78 it was up to 1000 by early '79, 100,000 by midsummer and a staggering 250,000 by the end of the year. Early 1980 saw 500,000 come and go and by the middle of that year there were an estimated 1 million pirates on 27MHz. Towards the end of 1980 the illicit CB market was said to be worth £1 million every month at retail prices.

This seemed to come as some sort of surprise to the government of the day, who promptly launched the first in a series of offensives designed to close the pirate network down for good. Their surprise is hard to comprehend, particularly in view of the formation of the Citizen's Band Association as far back as 1976. Dedicated to the legalization of CB by democratic means, the CBA, under the guidance of its vociferous President, James Bryant, had been directing a stream of communications to the Home Office and 10 Downing Street suggesting the prompt legalization of a CB

facility on a sensible frequency and hinting darkly at the disaster in store if this advice was not heeded.

The government response was to deny that a suitable frequency for CB existed and to prosecute as many pirates as it could catch. If only it had dedicated as much time, effort and money towards the research for a suitable CB facility as it did on the night raids and the patrolling Customs officers the whole thing would have been resolved quickly and easily. While groups of up to 15 Customs officers and policemen performed post-midnight raids on private houses, to confiscate illegal CB units one at a time, and then wrote smug, self-satisfied letters to their in-house magazine, *Portcullis*, the CBA and others were trying to find an answer to the problem.

Their first suggestion was based on the fact that 27MHz is, as we know, the poor end of the radio spectrum. They urged that a VHF, FM service be introduced at once. This would, they said, give us a first-class CB facility, avoid the problems of interference and skip, avoid the possibility that the country would be flooded with Japanese units and thus provide a shot in the arm for the ailing British electronics industry and also give us a head start in the world-wide search for a high-grade service to replace 27MHz. The frequency they proposed for this was 232MHz. 'Allocated,' said the Government. 'True,' said the CBA, 'but to the Lancaster bomber, which has been out of service since 1958.'

This stalemate on the question of frequency forced the authorities to come out of the closet at last. They finally admitted, in the early part of 1978, that in their opinion 'the disadvantages of such a service far outweigh the advantages'. At the time they did not see fit to disclose the facets which they saw to be disadvantages, although everybody with half a brain could see what the advantages were likely to be.

21

Although it was never mentioned at the time it was possible to draw some conclusions as to the reason for governmental opposition to CB. At the time the GPO (as it then was) had enjoyed, on behalf of the Government, a total monopoly (and therefore control) over all forms of communication in this country other than face to face conversation with one other person. Think about it. If you wish to communicate with anybody you must go through an official government agency or licensed body – telephone, postal service, TV, music radio, whatever. Even a meeting consisting of more than three people can be held to constitute Conspiracy or Riotous Assembly. It's a form of below-the-line control which any government would be reluctant to relinquish, and the introduction of a publicly accessible radio facility would mean just that.

For this reason the government fought hard against CB. At one stage they said they were afraid it would be used for 'non-serious purposes'. Apart from the fact that there seems to be little to complain about in that (and indeed, if it were to be used for *serious* purposes, like plotting the revolution, someone somewhere could well be in trouble) it was a statement which ignored the fact that via the medium of television and radio vast tracts of radio spectrum were already given over to purposes of unmitigated stupidity, never mind non-serious.

Further, it appeared, HMG did not 'take kindly to the idea of vast armies of people within easy communication of each other'. At the time many campaigners felt it wiser not to mention the existence of the phone service lest it be arbitrarily outlawed. The same Government spokesman who had already made these earth-shattering revelations, Lord Wells-Pestell (who probably wishes he could remain anonymous), went on to point out, while remarking on the disadvantages of CB in general, that he did not think 'that society takes very kindly to people who feel that they have liberty to rob, plunder, rape and do all sorts of things'. In this belief he was indubitably supported by every member of the House present at the time, but unfortunately My Lord neglected to indicate to his spellbound audience exactly how this opinion was related to the subject under discussion, to wit, Citizen's Band Radio.

Indeed the whole debate seemed to have little to do with CB at all, and several of the noble Lords present became quite entangled in a complex side-issue concerning the facility with which British

Rail had apparently lost two trains in a blizzard. The discussion ended inconclusively which, under the circumstances, comes as no surprise to anyone.

While the bickering continued the ranks of the pirates continued to swell. Their encouragement was coming from several quarters by now. In the mid-seventies the record *Convoy* had blasted its way into the history books, which was a considerable feat for such an undistinguished piece of inanity, but the CB jargon contained therein had fired the imagination of journalists all over Britain. As the number of prosecutions for illegal use of CB began to rise the media dedicated disproportionate amounts of space to the subject, principally in order to explain that having a reefer on didn't turn Cabover Pete into a dope fiend. The Laker walk-on service to the USA meant that there were hordes of people sampling the delights of CB in its natural habitat and smuggling it home for use in an unnatural one, and the collapse of the skateboard craze meant that there were hordes of out-of-work entrepreneurs all looking for a product to market. Suddenly they became CB experts overnight. In 1978 you had to know someone who knew someone whose cousin's friend's brother could get CB rigs. They cost about £100 for a straightforward 40-channel mobile. By 1979 you could buy rigs in just about any pub in the country and by 1980 the high streets were full of shops which never sold naughty, illegal CB rigs but which mysteriously had a huge market for the kinds of accessories which would be totally useless to anyone unless they owned a CB. Piracy, lawbreaking and commercial enterprise made CB very big business indeed.

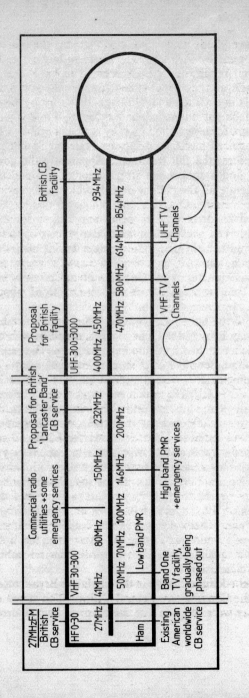

This simplified drawing shows just how crowded the radio spectrum is getting. And we haven't even put in Radio One, the marine and aircraft services, Diplomatic Radio and the multitude of others.

And although everybody made a point of watching the wall while the gentlemen went by they all knew where to buy a rig. They must have done, because everybody seemed to have one. And meanwhile, back at the CBA, James Bryant was still writing letters to the Home Office and Number 10. Only by now he wasn't saying 'look out' any more, he was saying 'I told you so'.

Public pressure for legalization was, by now, enormous. Clubs, magazines, organized marches, demos in Trafalgar Square, television programmes, the full treatment. Apparently the government couldn't find a spokesman stupid enough to stand up in public and reiterate all the clumsy denials which had previously formed their sole argument against CB. No longer was the Home Office prepared to deny as anti-social and useless something which had long been accepted as being the precise opposite in practically every other civilized country in the world. Instead they retreated to a lame variation on a very old theme, suggesting that CB was, after all, not such a bad thing, and wouldn't it be lovely if there were some room for it in our overcrowded airspace. Unfortunately, however . . .

This time it didn't wash. The pro-CB lobby had grown up. It had been a lovely baby but by now it was a monster. It had the support, in theory at least, of some 60 MPs, and the practical assistance of rather fewer. It had its own national body – The National Committee For The Legalization of Citizen's Band Radio, or NATCOLCIBAR, which was supposed to be an abbreviation. Most people just called it The National Committee. From now on the argument changed its tack rather dramatically; instead of various interested parties being forced into lengthy dissertations as to why we should have CB the question everybody was asking was – why shouldn't we? The Labour Government concluded that we shouldn't because it would interfere with existing services, could be used for criminal activity and would break the existing Post Office monopoly on communications. Only one of those statements, the latter, could be justified, as anyone with half a brain could work out. And since that appears to be the prime qualification it's a surprise that no politician arrived at that conclusion.

But there was a definite change in the air. James Bryant received a letter from the Conservative Central Office which stated quite clearly that they were prepared, in the event of being elected to office, to consider the question of CB in a favourable light. Probably this was not the sole reason for the dramatic Conservative victory in the election, but it cheered up the CB lobby no end.

Almost immediately the Greater London Council published a discussion document on the subject of CB, stating that if the majority of Londoners were in favour of it then they would support the campaign. If it was, as it later appeared, a market tester for the Government's own discussion document then it's hard to see why in the face of an overwhelming 'yes please' to 27 MHz Willie Whitelaw allowed his Green Paper to be published. Junior Home Office Minister Timothy Raison had guardedly remarked that the arguments in favour of CB, especially with regard to the freedom of the individual, were strong, but expressed doubts as to whether the service was practicable. In any case, he said, it would never be on 27MHz, which seemed like a pity, particularly when you consider that a very creditable service already existed on 27MHz and had every appearance of something that would be around for a long time, despite being illegal. It had long been too late for alternatives to 27MHz, but no one in Whitehall seemed to realize.

The Government proposals for CB finally appeared in August of 1980 and were met with scorn and derision the very next day in the national press and later in all the magazines concerned with CB, as well as by all the campaigning bodies. If it hadn't been so important and so sad it would have been hugely amusing. The official attitude was quite neatly summarized in the Home Office's total refusal to call the service CB. It was Open Channel and therefore not the same thing at all. In fact the only difference between what the campaigners wanted and what the government were offering was frequency, but that was more than enough. The Home Office had totally ignored all existing CB facilities, and totally ignored all the various demands – for 27, 41, 232, 450 and anything else. Instead they had gone straight to the UHF bands and 928MHz, having dismissed all other frequencies in a couple of sentences. It was quite clear that they had never considered the possibility of allowing 27MHz or anything else and were determined to saddle this country with a unique and useless frequency. By coincidence many other countries were examining this area of the radio spectrum, mostly with a view to utilizing it in addition to 27MHz, either as an extra CB band or for the more sophisticated PMR (Private Mobile Radio) sets. It may have been churl-ish, but there were many people who felt that the Home Office decision to opt for 928 was not motivated by any desire to provide this country with the best possible CB facility, but to give us a head start in what looked like becoming a fairly lucrative export market.

The Green Paper was hard to come by – it was necessary to write to the Home Office in order to acquire a copy – but even so the Home Office was inundated with 9000 letters from people wishing to protest against the proposals it contained. The princi-pal reason for the objection was not simply that existing pirates didn't like the number 928 and preferred 27 because it was easier to remember. The sad truth was that a CB facility operating on 928 was unlikely to provide a mobile-to-mobile range of more than a mile, if that, except under ideal circumstances in open country. Tests in Germany revealed that the urban range may be only a few hundred yards. If it had been introduced the term 'eyeball' would have disappeared for ever from the CB vocabulary because you would have had to be able to see someone for them to be able to receive your transmission. In fact the proposed Open Channel service would have reduced CB to being the toy suc-cessive governments had been so afraid of for so long, and ensured that its *only* use would have been for non-serious purposes.

The discussion document provoked exactly the kind of reaction the Home Office had been trying to avoid. On the one hand it stirred up a storm of protest which began, at last, to get some serious attention from the media and it also fuelled a kind of mini-boom in sales of illicit 27MHz sets. A great number of people had been holding back on purchase, waiting for the announcement which the Government had said was coming, and hoping they could get the legal feeling fairly swiftly. The Green Paper made it obvious that there was little hope of this happening and so they bought pirate sets instead. By the end of 1980 esti-mates of the number of people using illegal rigs had shot up to anything between one and two million.

During the course of that year CB was used, and reported in the press, for apprehending assorted criminals and also on several occasions during searches for missing children. In every instance this latter was with the full knowledge of the police in charge of the searches, but no prosecutions resulted. Increasingly, and especially in rural areas, it was apparent that CB had gained a widespread if unofficial acceptance. Back at the Home Office everything had gone quiet, but the rumours were buzzing like mad.

Finally, in February 1981, the Government were forced into the situation they had spent so long trying to avoid by any method other than the right one and which they had been warned about repeatedly since 1978, and it was announced that a CB service operating on 27MHz would be introduced in the autumn of the same year.

However, old habits die hard and even at this ridiculously late stage of the game the same old face-saving manoeuvres were being faithfully trotted out as it became clear that unlike the American system in widespread use the British system would be on FM, not AM. Although FM is a technically superior way of going about things it did seem rather futile at the time. Worse was yet to come, believe it or not, as the draft specifications of the service were issued, in photocopy form and on very restricted circulation one Wednesday in April. Written comments were required within 48 hours, but Good Friday was just 36 hours away – hardly long enough to read the specification, still less form a sensible opinion of it.

The reason for this steamroller attitude was immediately apparent; having acknowledged that the people wanted 27MHz to be in line with the rest of the world the Government was proposing a set of frequencies in the 27MHz band which were completely unique. Sets built to this specification would be useless beyond the frontiers of the UK and sets belonging to visiting foreign nationals would be rendered just as useless in this country. Further restrictions on antenna size, construction and installation virtually halved the permitted power output of the sets. Once again, while seeming to accede to public opinion the Government had in fact done the reverse, and saddled the country with a CB facility which was virtually useless and bore no resemblance to the service for which the campaigners had fought so long.

Despite frantic appeals from interested bodies within the UK and on the continent, and despite a meeting between members of NATCOLCIBAR and William Whitelaw himself, following which the Home Secretary could have been in no doubt whatsoever that what he was doing went against the wishes of everyone except the

Home Office, the specification for the new service was published in June 1981.

It contained no reference to licence conditions or fees nor to any date on which the service would be legal. Neither did it lift the ban on the manufacture of 27MHz equipment imposed in 1968, or even indicate when this ban would be lifted. It did, however, ensure that in factories all over the Eastern hemisphere little yellow fingers would be busy making CB rigs to the new British specifications, against the day when the import ban on 27MHz equipment should be lifted. After which it will be CB for all, with the exception of British industry.

In the end the manufacturing ban was lifted at the beginning of September. In effect this didn't mean that hundreds of Brit electronics firms started building CB rigs. What they did was to start importing the bits from the Far East and start screwing them together. It wasn't until the very beginning of October that the Home Office announced the actual date for legalization of the new FM services – November 2nd – and the licence fee. This turned out to be a not altogether unreasonable £10 to cover up to three sets operating on either of the new frequencies.

From then on the service was in action, and only time will tell whether the British public will accept it, or reject it in favour of the still-illegal American system as so many of their continental counterparts have done.

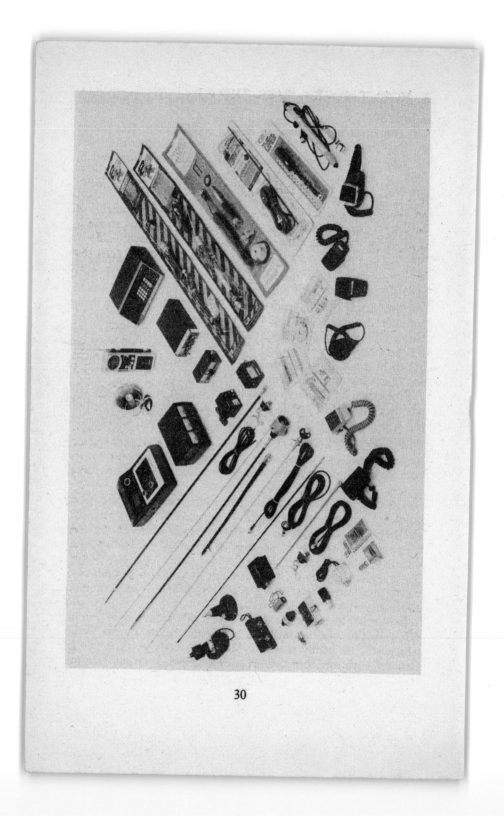

CB radio is, fortunately for most of us, an extremely simple thing. It is only fractionally different to the music radio fitted to most production cars as standard at the factory in that it is capable of both receiving and transmitting radio signals, for which reason it is properly known as a transceiver. Because the transmit side of the unit is somewhat more sensitive to bad treatment than the receive side, the type of antenna required and the way it is matched to the set is somewhat more critical than you're used to, but apart from that there are no problems to be overcome.

All a radio transmitter does is to generate from its antenna an electrical signal which vibrates, or oscillates. The number of times which the signal oscillates every second is measured in Hertz and is the frequency of the signal. Thus a 10Hertz signal would be of lower frequency than a 100Hertz signal. Both would probably kill you, which is why we tend to use higher frequencies. CB operates on 27 MegaHertz, and since Mega means million you can easily work out for yourself that a CB signal oscillates at the almost incomprehensible rate of 27 million times every second. Fortunately this staggeringly high figure is not incomprehensible to a receiver tuned to the same frequency, a fact which has contributed greatly to the success of radio transmissions in our modern society.

Radio signals have often been likened to the ripples made in a pond by a stone – they spread out in all directions and are regularly spaced. The distance between each crest and the one before it is the same as the distance between the one behind, and the one behind that, and the one behind, and . . . get the picture? Because the distance is always constant it may also be used as a means of reference – wavelength. 27MHz CB has a wavelength of about 11 metres.

Even more interesting is the discovery that radio waves, seen in side elevation, are symmetrical above and below a centre line, which will be handy later on. Meanwhile let's go back to the Carrier Wave, which is what we call a simple radio transmission

A carrier wave, such as you might create simply by holding the PTT bar constantly open.

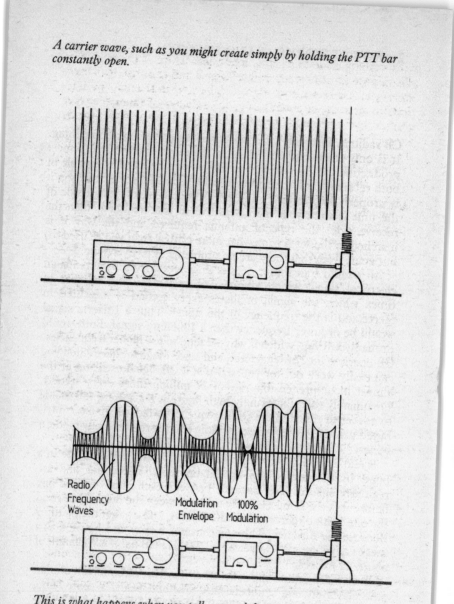

Radio
Frequency
Waves

Modulation
Envelope

100%
Modulation

This is what happens when you talk, or modulate; in fact the radio waves make a 'picture' of your voice. Not a pretty sight.

with nothing done to it. By altering the nature of the carrier we can impose on it degrees of intelligence and thus use it to carry messages. There are two ways of achieving this; either by fractionally altering, or modulating, the number of times the wave vibrates every second – Frequency Modulation (or FM) – or by altering the height of the wave above and below the central line – Amplitude Modulation (or AM). In both cases the radio wave still retains its symmetry about the centre line. The most obvious use of this phenomenon is Morse code, in which a simple series of short or long dashes can be sent with ease.

The use of a microphone, which converts the vibrations of speech into electrical impulses, allows us to impose the patterns of the human voice onto the wave and allows a receiver to convert it back into vibrations and then audible speech through a loudspeaker. Which is everything you need to know about how radio sets work. Probably in this case pictures are worth ten thousand words apiece, so they're worth looking at for a moment or three until you're absolutely certain you're with us.

Good.

Now, American-style CB uses AM to transmit speech from one place to another, but we have chosen FM in this country. You could be forgiven for thinking that there's no special reason for this but, on a technical level at least, you'd be wrong. AM makes the carrier wave get bigger or smaller or both in rapid succession. Getting smaller is fine, but when it gets bigger we start to have problems. If there were only two radio sets in the world this wouldn't matter at all, but there are sadly rather more than that. In fact Britain has the most overcrowded airwaves in the world, worse even than New York. Nip out and have a look at the sky. It might look like just sky to you, but it's actually full of messages zipping about from place to place. Some of them may even be to your advantage. It makes sense, therefore, to make sure that they all operate at different frequencies so that they don't get mixed up, otherwise you could find that your TV set gets confused and starts showing the closing moments of the last Test to the sound accompaniment from the local taxi firm. Bad news for you and bad news for the bloke who wants to get to Neasden but can't get a cab because the driver's transfixed with excitement in a lay-by listening to the last two overs. In order to minimize the possibility of this occurring HMG have made the dispensation of radio

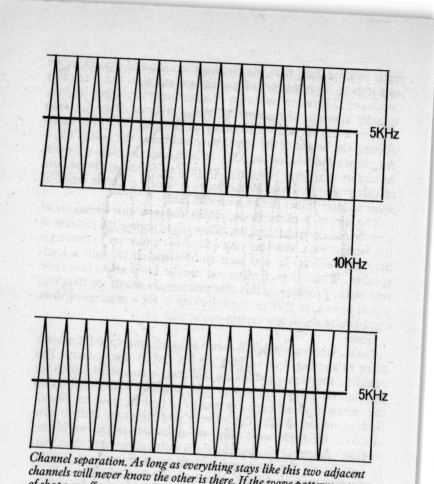

5KHz

10KHz

5KHz

Channel separation. As long as everything stays like this two adjacent channels will never know the other is there. If the wave patterns get out of shape or off-centre you start to get bleeding into the nearby channels.

spectrum the prerogative of the Home Office. It is their job to make sure that all users are separated from each other by a distance which is sufficient to prevent any mix-ups.

CB uses 40 channels, and it is considered that 10KHz of space between each channel is sufficient to prevent bleeding, or cross-talk. This allows a total of 10KHz for modulation (5KHz each side of the centre line) before the channel crosses into those either side of it. Under normal circumstances this should be sufficient. However since AM signals get bigger and smaller there is far

This is a carrier wave, and is dead boring, but at least be grateful that it's there. If it wasn't you'd never be able to speak to anyone.

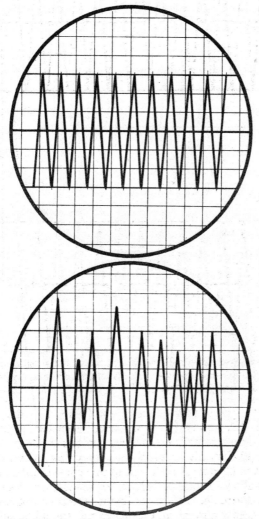

Simple modulation, caused by keying the mike, just like in Morse code; you can make long dashes or short dots in the wave pattern.

Voice modulation of the carrier wave. This is AM, and you can see that the wave pattern gets taller and shorter, growing out of its allocated bandwidth towards nearby channels . . .

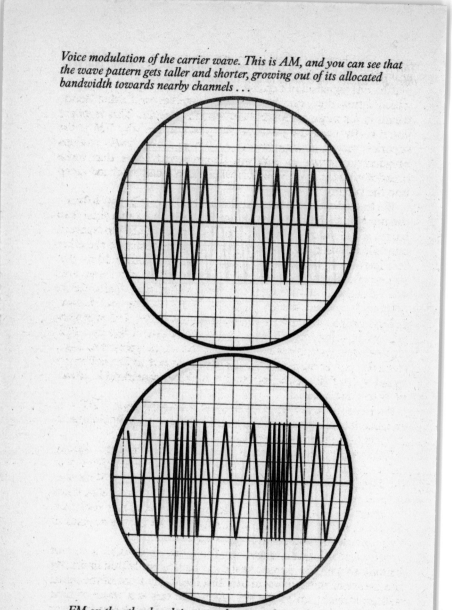

. . . FM on the other hand, just gets closer together or further apart; clearly this is less likely to cause interference than AM.

more likelihood of them creeping into nearby channels and annoying the neighbour than if they stayed within their bandwidth. FM signals don't change size; since it is only the frequency which alters they simply get closer together or further apart, which is far more acceptable and far, far cleaner. This is where you'll really need the pictures. But you can see why FM is the superior means of transmission. Even in mono, radio stations broadcasting music on FM will always sound better than those broadcasting on AM. Which is where stereo and sideband creep into the picture.

We have discovered that the radio signal comes in two halves – the one above the line and the one below it – and that these two halves are identical. Which means that you only need to transmit one half in order for a suitable receiver to manufacture the other half itself and produce an intelligible result. If you could do that you would be able to have two channels where previously only one existed and you would have invented Sideband. Quite clearly Sideband can be divided into two bits – Upper and Lower, depending on which half of the signal you use, and is a very wonderful thing indeed, since it turns 40-channel rigs into 120-channel devices and makes far better use of the spectrum. Also, since the power output of the rig is compressed, it can deliver its signal up to twice as far as usual, which has considerable attractions to many users.

And the attraction of sending over a long distance – DX – is enhanced by a property which, although not actually unique to 27MHz CB is not found at higher frequencies. Skip.

This magic phenomenon is peculiar to radio signals working below about 35MHz and has been the cause of much concern to the Home Office when deciding on the nature of our CB service. A great number of people believe that the odd channel allocations for British CB were made specifically so that legal sets could not be used for DX work simply because they do not correspond to frequencies anywhere else in the world.

Most radio signals spread out from the antenna in a sort of balloon shape. The precise shape can be altered, within limits, by the nature of the antenna in use. But inevitably some of the signal will go straight up in the sky. As a rule this is a waste of time unless a great number of your friends are pilots or live in tower blocks, which is why most antennas are designed to concentrate their power into the groundwaves (which spread out horizontally) rather than the airwaves. Most airwave signals just keep on going up, straight off into space. Below 35MHz a peculiar affliction strikes them; the ionosphere reflects them back to earth. This was how the existence of the ionosphere, of which we had previously been ignorant, was discovered.

Once reflected these radio waves need not return to earth at the point from which they departed. In fact they are more likely to turn up somewhere completely different, where they may be totally unwanted, very possibly thousands of miles from their birthplace. This is skip, and it can allow communication between this country and America or other exotic locations.

Skip is not something which can be counted on, however, and varies according to the time of day and the weather conditions, as well as the faintly nomadic inclinations of the ionosphere itself. Sunspots, believe it or not, and you'd better because it's true, also affect skip. It gets worse when they do. Sunspot activity moves roughly in an 11-year cycle and won't reach its next peak until about 1992. The last one was in 1980, which accounted for all the free Italian lessons we were getting every afternoon in the summer.

As far as the really keen DX workers are concerned sunspot activity is much too chancy to be relied on for their purposes and very often they are prone to giving it a little help along the way. This help generally takes the form of a highly illegal device known as a linear amplifier. It is cheap, simple, easy to install and has dramatic returns in range increase for fairly minimal increments in power. A linear amp may boost the power of a 4-Watt CB up to anything between 25 and 250 Watts. At those upper power levels skip becomes redundant anyway, because from a high enough hilltop the signal the rig puts out will go right round the world and hit you in the ear from behind.

That kind of signal strength, emanating from somewhere in Hackney, will have highly desirable properties for people living in Brighton and points south, but will not be so funny for the poor chaps in East Ham. The effect on nearby CB rigs will not actually be completely similar to the detonation of a thermonuclear device under the dashboard but nevertheless it can be extremely un-healthy. For this reason people with big boots (as the jargon goes) are not always terribly popular with other breakers. Neither are they an overwhelming success with the local population, who may resent their TV screens imitating the action of a washing machine.

The problem of TVI was always at its highest in the States during the currency of the early 23-channel rigs; later improvements in technology and manufacturing standards and the higher specifications required for the newer 40-channel sets have cured the problem to a large extent, and the fitting of filters to the CB unit or the TV or both should prevent it completely, unless the CB rig is very close or banging out two million watts. The trouble here is harmonics, which is nothing to do with Larry Adler or the Edwin Hawkins Singers.

All radio sets are prone to both giving and receiving interference; this is sad but true. In the main it is proximity which makes it happen. If you park a radio transmitter close enough to a radio receiver then the receiver will receive, even if the two units are operating on different frequencies. You've probably experienced this already at a basic level with your TV or when tuned to your local FM station. More often than not it's a member of the local taxi firm who announces collection or delivery of a passenger at the house next door to you whose message will tramp all over *Blankety-Blank* at a particularly critical moment. It need not be a taxi, of course; any radio set is capable of it, whether it belongs to the fire-brigade, police or Gas Board. Or it could be the CB rig in the jacked-up Cortina next door. The question is, how does a radio set working on, say, 85MHz, interfere with a TV set working on, say, 850MHz? Even if it is very close?

Harmonics, that's how. These little fellows are simply multiples of frequency and are dead easy to work out for anyone who got O-level Sums at school. The rest of us will need pocket calculators. The first harmonic of 27MHz is 27MHz. The second, multiplied by two, is 54MHz. The third, multiplied by three, is 81MHz and the fourth, obviously enough, 108MHz. The thing about radio signals is that they are dead keen on mathematical relationships, which is why quarter-wave antennas work as well as they do. So a 27MHz signal is half the size of 54MHz which is bad news for anyone operating 54 with a half-wave antenna. Also why the early Stateside rigs, working 27MHz, interfered so much with the TV broadcasts which were (and often still are) in the same part of the radio spectrum as our old Band One TV over here, and not terribly unadjacent to 54MHz. Follow it further, and the third harmonic of 27 lands squarely into the low-band PMR frequencies while the fourth drops it on the head of the emergency services band, mostly police.

The Home Office claim that this is another reason why the set of channel allocations they have made for British CB are so odd; they say that the individual channel frequencies have been carefully worked out to minimize the risk of this occurring.

CB operates in the 27MHz band, using 40 channels. Every one of those channels is an exact frequency close to 27MHz separated from its neighbours by 10KHz. You will see that it is different from the American spacing, and also that the frequency of the London area calling channel, 14, is not 27MHz precisely but in fact 27.73125MHz. Using the well-known pocket calculator again it is evident that the harmonics of channel 14 are 55.4625, 83.19375 and 110.925MHz which, according to the Home Office gnomes, is no accident, and puts the potential interference as far as possible into empty space. Hmmm. In view of the fact that the channel spacings are so regular this seems rather unlikely in every case, but hardly worth worrying about.

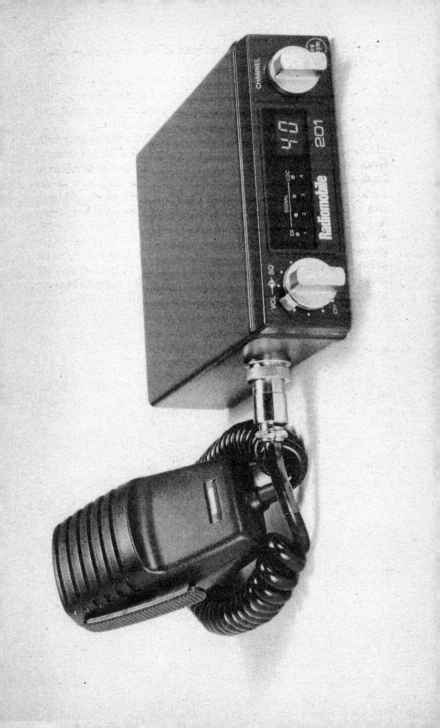

What will it mean to you? The answer to that question is not very much at all unless you are already a dedicated follower of AM, in which case they will probably have driven you to heights of unequalled rage already.

The British Specification (MPT 1320) is printed in full elsewhere for the benefit of anybody sufficiently interested to read it, but in truth it is of little interest to the average breaker; it is a set of design specifications which enable manufacturers to produce the equipment as the Home Office have chosen it. All equipment which conforms with this standard is required to carry a permanent mark – CB 27/81 – on the case. Any CB equipment which does not have this mark, or which carries it only in sticker form does not comply to the standard, is not eligible for licensing and its user may well suffer confiscation and prosecution. If it does carry the approved stamp you won't get prosecuted, but you probably won't get much else either.

The specification provides for 40 channels, with their carrier frequencies spaced at 10KHz intervals and a maximum power output from the transmitter of four Watts. As a brief generalization this sounds not totally unlike the American system in widespread use by the pirates in this country and in legal use in about 60 countries around the world. Closer examination reveals a wealth of discrepancy, however, some of which will not make itself felt for some years to come.

To begin with the only type of modulation permitted is FM, which in itself is no bad thing on a purely technical level as we have already seen. FM gives a clearer, sharper signal, and causes less interference to others. Fine. However, it is not only transmitter power which affects radio transmission. Aside from the addition of a linear amplifier to boost power there are other ways in which better performance can be obtained from a transmitter. The most obvious of these is the antenna itself.

Under ideal conditions an antenna would work best if it was exactly the same size as the wavelength being operated. In the case of CB, on 11 metres, this would give us a vertical pole somewhat in excess of 30 feet high. This is clearly impractical even for base station use and totally out of the question for mobile purposes. Luckily for everyone, radio signals are more than happy with an antenna which bears a mathematically exact relationship to the wavelength; half-wave, quarter-wave and so on. This is still not altogether good news for mobile users, since a quarter-wave antenna for 27MHz will still be about nine feet high and will therefore still be in a position to argue with bridges and overhead power lines. All is not lost, though, because antennas are basically a bit dim, and it is sufficient for them to believe they are nine feet high for them to behave as if they were. What happens, then, is that we take an antenna three feet high and attach to it a piece of wire six feet long wound round and round very tightly into a coil only a few inches long. The provision of such a loading coil gives an antenna an electrical length of nine feet while limiting its physical length to three feet and four inches. The result of this is a highly practical device which works perfectly well and also fits into multi-storey car parks with no problems. The loading coil may be added at the bottom of the mast (base-loaded) at the middle (centre-loaded) or at the top (quite right). Top-loaded antennas work better than either of the other kinds.

As far as MPT 1320 is concerned antenna height is limited to a maximum of 59 inches, which is longer than most people want on their car and shorter than quarter-wave anyway; it will need a loading coil. This may only be placed at the base of the mast, in the most inefficient of the three alternative locations. Worse still, the antenna must be a simple wire or rod, with none of the helical tuning which is a commonplace among the antennas in use by AM breakers all over the world. This will reduce the efficiency of the antenna still further, both in transmit and receive modes, critically reducing the sensitivity of the unit to which it is attached.

An antenna need not only be used to improve performance; it can also be used to restrict it. In the end the power of the transmitter is irrelevant. What actually counts is the Effective Radiated Power (ERP) from the antenna. MPT 1320 allows four Watts transmitter output but only two Watts ERP. The specifications actually provide for a service which will be only half as efficient as the existing pirate AM network. Neither are the restrictions confined only to mobile use. Let's go back to base,

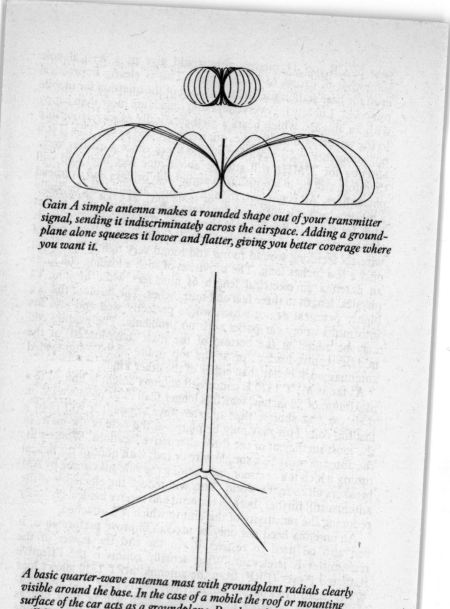

Gain A simple antenna makes a rounded shape out of your transmitter signal, sending it indiscriminately across the airspace. Adding a ground-plane alone squeezes it lower and flatter, giving you better coverage where you want it.

A basic quarter-wave antenna mast with groundplant radials clearly visible around the base. In the case of a mobile the roof or mounting surface of the car acts as a groundplane. People with glassfibre caravans will find a flat sheet of metal very useful for sticking a magmount onto and creating a groundplane effect.

also to the philosophy whereby your antenna may be used to advantage rather than disadvantage.

Up until MPT 1320 the most basic form of antenna for fixed use was the quarter-wave. Just a straight rod nine feet high. Improvements to the performance of this antenna can be made by adding small radials around its base, whereupon it becomes a ground-wave antenna. All this does in effect is to squash the balloon of signals flatter, thus sending them further along the ground and not so far up. The increase thus achieved is called simply gain and is measured in decibels (db). It's not altogether straightforward unfortunately, but it's not that difficult either.

Apart from the basic ground-wave it is possible to complicate antenna design still further by adding extra elements, so that it looks like a TV aerial. Only one of the elements in your antenna will be connected, via the central core of the co-ax to the rig. This is called the driven element. The others (let's keep it basic and only have three elements for now) simply give it a little help. Behind the driven element is the slightly longer reflector and in front is the shorter director. These extra elements all combine to treat your radio signal in much the same way as a car headlight treats light. The source of light is the bulb, which corresponds to the driven element, and sends its signal in all directions indiscriminately and none especially well. The reflector in the headlamp is the reflector element of the antenna, which compresses the signal into a vague sort of beam and ushers it in a single vague direction – forwards. The third, director, element behaves just like the lens; it concentrates, narrows and focuses the signal, sending it powerfully in one direction. All the signal energy which was previously being dissipated about the place is travelling one way only now, and will consequently travel further in one direction than it would in any of the others had it been left to its own devices. Quite clearly the gain from this sort of antenna – a directional beam – is considerable. It also requires a great deal of thought from the operator, since he or she will need to know in advance in which direction the signal should be travelling. Either that or it requires some extra cash outlay in order to acquire a rotator. Basically a simple electric motor operated by remote control, this is mounted at the top of the mast and can then be made to aim the beam towards any point of the compass.

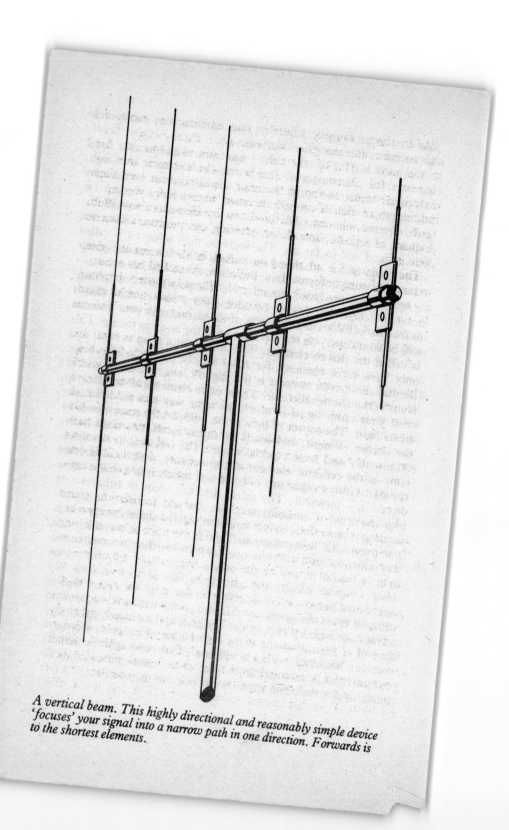

A vertical beam. This highly directional and reasonably simple device 'focuses' your signal into a narrow path in one direction. Forwards is to the shortest elements.

Adding extra elements, creating stacked beams, or quads will also increase antenna gain. For example a three-element beam should have a gain of about 8db, four-element 11db and five-element 14db. Making a stacked beam – having more than one directional beam on top of the mast – will produce even more impressive results. Two three-element beams hoist you up to 12db, two four-elements 14db and two five-elements about 17db. A quad, as a simple five-element device, can produce a massive 20db gain.

The penalty for all this gain comes in the increasing size, weight and complexity of the installation. A stacked beam carrying two five-element loads at each end will need a powerful rotator to turn it and the sort of carrying mast which would make Isambard Kingdom Brunel more than slightly envious. This is really only for the very serious-minded.

Also be advised that gain is not a simple multiplication sum. An antenna with 10db gain is not twice as good as one with 5db gain, although it is obviously an improvement. For the purposes of calculation a basic quarter-wave is considered to have no gain whatsoever, and the performance of other antennas is measured against this. An antenna with 3db gain will have the same effect as running an 8-Watt transmitter through a straightforward quarter-wave as opposed to the permitted 4-Watt, and is thus twice as good. A five-element Quad, with a gain figure of up to 20db has the same effect as running 200 Watts through a simple quarter-wave.

For base stations, then, a proper antenna installation is not altogether dissimilar in effect to increasing the transmitter power. Presumably for this reason MPT 1320 does not allow anything other than a 59-inch vertical rod, which prevents anyone from enjoying the benefits of antenna technology. And it gets worse.

Any antenna will produce better results as its height above ground level increases. Ask yourself why your TV aerial works better on the roof. So when you install your antenna you will want to put it as high up as it can safely be accommodated; either on the roof of your house or at the top of a very long pole. More so since you are stuck with a simple rod. Bad news again is that if your antenna is mounted more than seven metres (about 23 feet) above the ground then your transmitter power must be reduced by 10db. This may not sound very much, 10db, but it will give

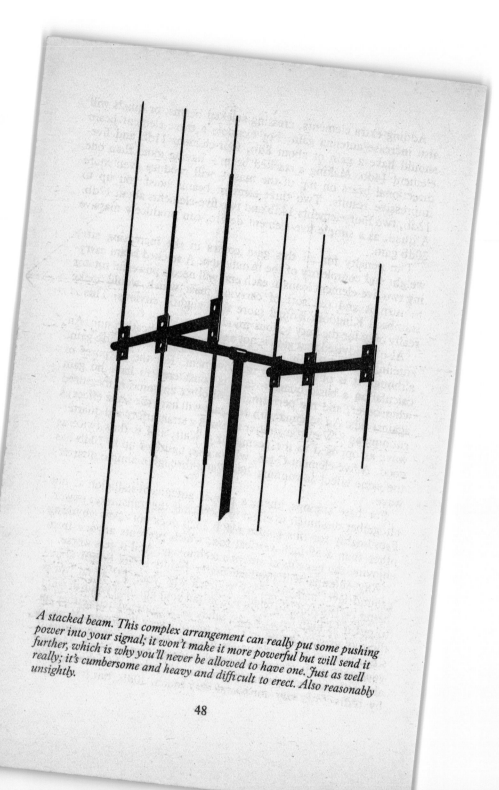

A stacked beam. This complex arrangement can really put some pushing power into your signal; it won't make it more powerful but will send it further, which is why you'll never be allowed to have one. Just as well really; it's cumbersome and heavy and difficult to erect. Also reasonably unsightly.

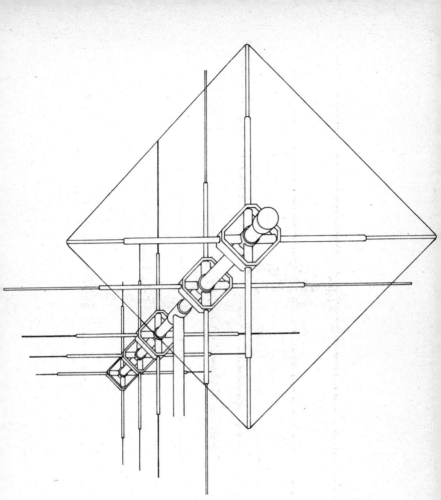

Now here is a highly complex device. A directional beam, turned by a rotator, it is also switchable between vertical or horizontal polarisation; if the elements are vertical it is vertically polarised, if they're horizontal it's horizontally polarised. Not much difference, except that radio waves from a vertically polarised antenna travel up and down and horizontal jobs swish from side to side. Inter-communication is possible but not good. So if you live near a motorway (all mobile antennas are vertically polarised) and you only want to speak with the betting shop down the road then both of you should get horizontally polarised. That way you'll hear less of the chatter from the superslab.

49

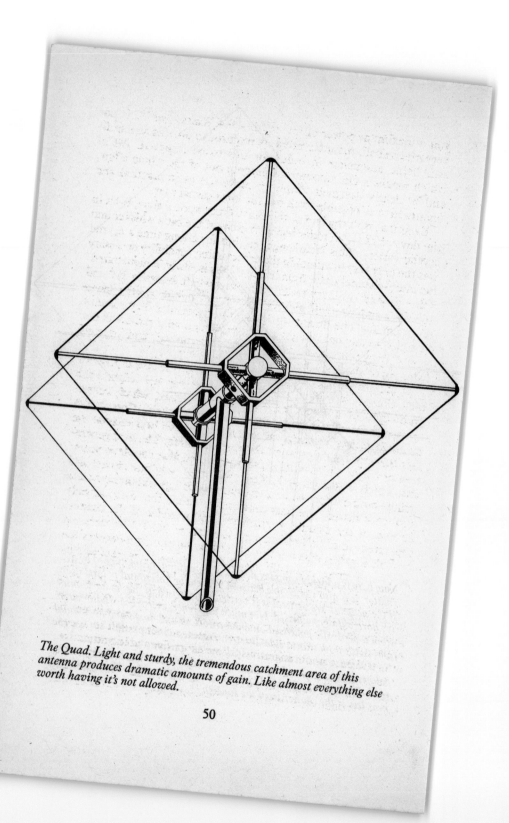

The Quad. *Light and sturdy, the tremendous catchment area of this antenna produces dramatic amounts of gain. Like almost everything else worth having it's not allowed.*

50

you a maximum power of one tenth – 0.4 Watts – which is not very much at all. Manufacturers are required to provide an inbuilt switchable attenuator in order for this to be achieved. All of which makes a fair amount of nonsense out of the whole affair, and is clearly designed simply to make CB as impractical and unattractive as possible to all but the most casual user.

Even that isn't the end of the story. Once upon a time, back in the days of 2LO, all radio receivers were of the cat's whisker and crystal variety. This is no longer so, but for a long time a crystal was the only way of attaining the exact tuning capability necessary for multi-channel radio transceivers. This is easily demonstrated. Tune in your ordinary radio set to somewhere between 100 and 105MHz and you will pick up all sorts of public utilities using multi-channel equipment. Although the crystal sets they use may be tuned to any one of a dozen or so channels your radio will not be sensitive enough to sort them out and you will find yourself listening to about six different conversations.

At one time it was necessary for a CB rig to have a crystal for each channel it worked on, but the microchip has allowed the production of a device known as a synthesiser, which cannot replace them all but does away with most of them. Current CB technology uses a system called mixer synthesis to overcome the fact that current chips won't operate at 27MHz. The next generation of chips, which will be available in the next couple of years, and which will work at 27MHz, will need only one crystal and will be both cheaper and more accurate. CB rigs will thus cost less and work better. The only drawback to this is that they can only be used if the channel frequencies are multiples of the channel separation or of half the channel separation. MPT 1320 does not provide for this, which will make utilization of the new generation of silicon chips virtually impossible. Since it is exceedingly unlikely that the Home Office boffins are unaware of this forthcoming development the only conclusion possible is that they have deliberately set out to make British CB as expensive as possible in order to deter people from using it. This will not be apparent yet, but as the cheaper American sets begin to appear ours will become comparatively more expensive and less attractive.

It is also conceivable that the frequencies were designed to annoy current Sideband pirates in the UK. Most DX and skip talking is done on Sideband and these operators, by virtue of their more expensive and thus less common sets, have enjoyed a fairly quiet and private piece of airspace for some time. The new specifications for CB will interfere with that quite successfully; already manufacturers testing sets to the new specification are finding themselves plagued by Sidebanders who are currently intrigued by the novelty of it all; that happy state can hardly be expected to last.

Whatever the reasoning, the end result can only be the same. In response to a huge demand for a workable CB service the Home Office have done their very best to provide something which is nothing of the sort in the hope, presumably, that it will all go away. What they have achieved is regrettably the precise opposite. Many millions of people, new to the concept and capability of CB will quite happily adopt the British system and doubtless extract a great deal of value and pleasure from it; there is no doubt that it will improve the quality of life for many. Likewise a great number of people who are already using the illicit AM sets will just boycott the new service completely, and carry on in their own illegal way. This situation, with two separate CB networks, will successfully ensure that neither of them will ever be able to realize their full potential; the major benefits from CB begin when there are a large number of users, and with some on AM and others on FM this can never happen to either properly. Those people on FM will never break the law and go AM, and those on AM will never abandon their service for one which is less than half as good as it should be.

In the end it will be a matter of individual choice; it may even be a question of compromise, as it seems likely that large numbers of breakers will use the new sets but ignore the antenna restrictions entirely. If this is the case then it may not be all bad, for it may persuade the Home Office to abandon its restrictions in this area, and simply doing this would be a marked improvement in the kind of service we could expect from British CB.

Sadly, although it will never be too late to lift restrictions on antenna type, it is already too late to alter the channel allocation, which means from now on the idea of commonality with the rest of the world goes out of the window, and on channel at least, we're all alone.

Of course you do also have the option of using the CB facility granted in MPT 1321 – 934MHz, which will allow you 20 channels

52

with a maximum ERP of 25 Watts. This should provide for at least five miles mobile-to-mobile range in open country and, since UHF signals are far more easily deflected by interposed solid objects (like trees and houses) than lower frequencies, probably about 800 yards in a crowded urban environment. Often you may be able to shout louder, and it is to be hoped that this type of equipment will necessarily be fitted with the PA option common on 27MHz equipment, since it will no longer be a luxury but a necessity, capable of much greater range than the transmitter itself. All this can be yours for a sum probably not very much larger than £300. Then again, you could stick to 27MHz for fifty quid. . . .

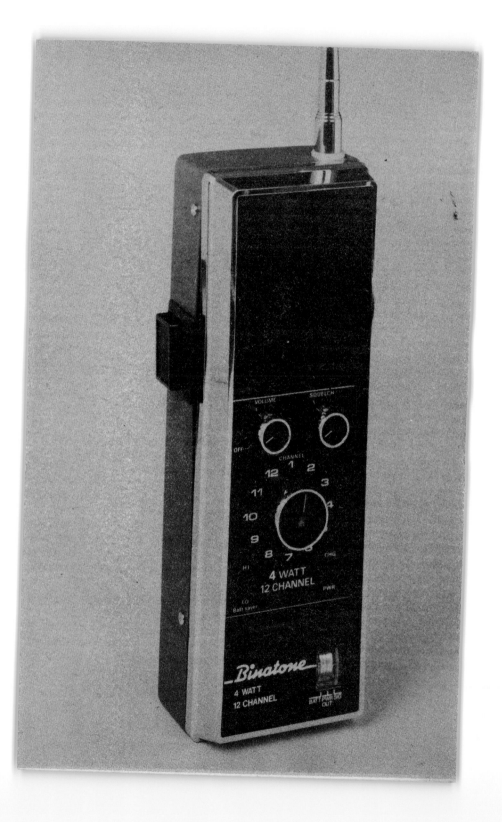

4   CHOOSING A RIG – FACEPLATE CONTROLS AND FUNCTIONS – HOME BASES

The whole object of CB is that it is a simple device which can be (and often is) operated by children. Unlike Amateur Radio it requires no specialized knowledge and should be no more complex in appearance or use than the average TV set and probably a whole lot easier than a video recorder. And just like either of these two devices it is not necessary for the operator to understand anything about the way it works in order to be capable of success at the controls. You don't need to know why your TV changes channels at the touch of a button, or why the vertical hold makes the picture wobble; it is sufficient that you are aware that it does. By the same token it is no part of your needs to understand why the squelch control on your rig makes the background static louder or quieter; as long as you are aware that it does you will be able to operate the rig successfully. This philosophy is often disparagingly referred to by some people as black box technology. Usually these are the buffs, the enthusiasts. Probably they've got a hi-fi set-up at home which resembles the flight deck of *Columbia*. Probably it is no more effective within the confines of the average living-room than a straightforward music centre. But it looks jolly impressive and provides a centrepiece for much informed (and uninformed) eloquence, even though it may be of no consequence to anyone at all if they can hear the conductor breathing during performances of the Ninth. As a rule they probably can't.

The same principle applies to a great number of CB rigs currently available in the States. They can sometimes look like the communications console on the *USS Enterprise*, but as a rule it is impossible to raise Starfleet Command when you need them, and shouting 'beam me aboard Scotty' at the top of your voice in response to a well-measured 'may I see your driving licence, sir?' generally fails to elicit any kind of response whatsoever. By much the same token many of the most impressive-looking rigs are incapable of any feats greater than those performed by the most basic unit.

The truth of the matter is that, even more so with CB rigs than with domestic hi-fi, the units are all made to a basic minimum specification and it is illegal for the performance of any of them to fall below that minimum.

Equally it is illegal for the performance of any of them to rise above a certain level. This puts you, as a potential purchaser, in a rather enviable situation. No one, in any shop, can give you a hard-sell line about how wonderful their products are compared to any other. By law they must all be the same within limits. All you have to do is check the stamp, make sure that the set does in fact comply with the rules, and part with the correct amount of green and folding. Nearly all high street multiples will be selling CB equipment soon – Dixon's, Smiths, Woollies, the works – which simplifies it even further, since if you're buying from someone like that you won't even have to worry about legality; it's more than any of their reputations are worth to unload duff gear on the punters, so you know you're going to be all right.

Having said all that let's go back to the Mission Control principle, as applied to hi-fi. Nine times out of ten appearance makes more impression on people than performance. A nice sleek satin-polished case with lots of switches and flickering lights will be a bigger seller than a dowdy old Quad unit, but I know which I'd rather have. People like, and will pay for, something which looks good, even if it isn't any better than the ugly stuff. And when the law says it can't be, better appearance will be the only possible selling point.

You may rest assured, therefore, that there are going to be cheap CB sets and expensive CB sets. In almost all cases the difference in price will be accounted for by a lot of cosmetic garbage which will have little or no value when you're in Chelsea trying to raise someone in Wembley. Some of the gadgetry, which sooner or later will inevitably include a digital clock, may actually make your life easier, may make your set simpler to operate and easier to maintain in top-class nick, and you may well feel that this is worth paying for; you may well be right. None of it will give you better range and performance, in which case you may feel it isn't worth paying for. The choice is yours, and since it's about the only one you've got make sure you have fun making it . . .

Meanwhile, let's have a quick look round the dials and knobs and make sure we all know what they're for. To be honest there

56

are only a couple which are actually vital to the cause, and after that it's just window dressing. You will find that handsets are less well-equipped than mobiles and mobiles less than base units. As a rule this is because handsets need to be small and light while having storage space for internal batteries (and often tend only to have limited choice of channels – single or triple is most common) and users of mobile units have to drive cars as well as operate their CB if they are to avoid the necessity for channel 9 broadcasts. Base stations have no such limitations and will therefore always be better-equipped.

On/Off/Volume

Easiest one of the lot. This works just like the switch on your music radio/TV set/deaf aid. Turn it clockwise until it clicks – you're switched on. Keep turning it and everything starts to get louder. Reversing the procedure reverses the effect. It may be worth pointing out that the volume control will only ever affect the strength of incoming signals and has no effect whatever on your transmissions.

Channel Select

This may come in several forms. Most common is a 40-position switch which will allow you to select any channel you desire by turning it in the correct direction. In this chips-with-everything age it is virtually certain that the channel you are tuned to will be displayed in living technicolour via an LCD readout placed conveniently adjacent to the control switch. In some cases it is possible that channel selection is achieved by pressing one of two buttons. These will be marked 'up' and 'down'. As a rule the set will automatically be on channel 01 when switched on and it will be necessary to press 'up' if you require channel 14, releasing the button at the appropriate moment. On the other hand if you want channel 32 it will be quicker to press 'down', since the unit is quite capable of counting backwards.

Squelch

Quite my favourite control, and very aptly named. It's an unfortunate truism that even on the very wonderful FM service radio sets always collect a large amount of spurious noise and interference. 'Hash', as it's often known. This can quite easily

57

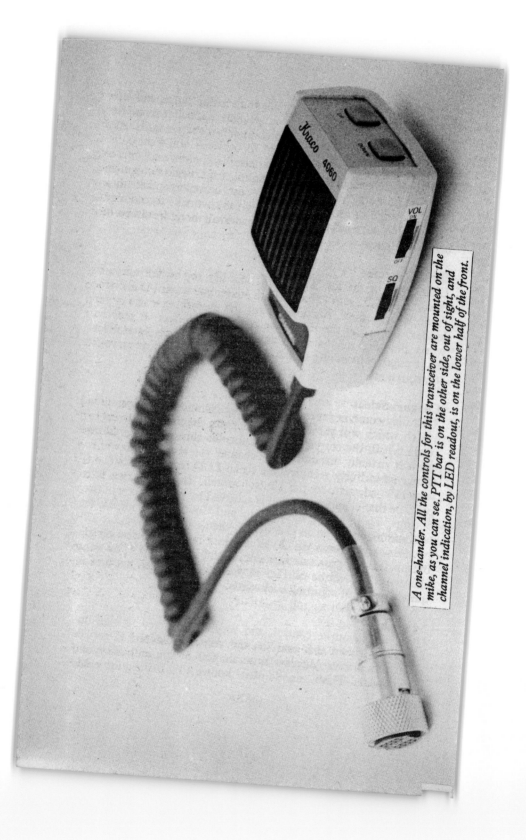

A one-hander. All the controls for this transceiver are mounted on the mike, as you can see. PTT bar is on the other side, out of sight, and channel indication, by LED readout, is on the lower half of the front.

make your teeth itch after a while and turning the squelch down (usually anticlockwise) makes it go away. Unfortunately it also makes everything else go away, as it reduces the sensitivity of the receiver, so you need to be careful. The trick is to find the point at which the background rubbish has just disappeared and then turn the squelch back up by a tiny fraction. At this setting silence should be golden, etc., but you'll also be able to pick up any transmission which is strong enough to give you an intelligible signal. If you're talking to a mobile moving away from you then you will need to open the squelch as the range increases.

Push-to-Talk (PTT)

This will be on the microphone, usually in the form of a bar pivoted at one end. Reasonably enough you must push it when you wish to speak and it will open the transmit side of your rig. Also remember to let go of it when you've had your say otherwise you'll never hear anything and neither will anyone else; CB rigs, like most two-way radios, are so constructed that it is impossible to transmit and receive at the same moment.

Automatic Noise Limiter (Noise Blanker) ANL (NB)

This little number is either off or on. Off, it does nothing. On, it has a fairly marked effect on incoming interference simply by cutting off the very top audio frequencies, since these are the ones which contain the bulk of the noise. It takes the hiss out of listening just like a Dolby on a cassette player. It shouldn't, if it's a good one, affect the sensitivity of your set to the same extent that squelch does.

Distant/Local

Almost a two-way squelch. In 'Local' it reduces the sensitivity of the set and prevents nearby transmitters overdriving it giving you a clearer, distortion- and interference-free signal, which is exceptionally useful if you're in convoy. Switching back to 'local' restores the set to full sensitivity and allows faraway breakers back into your life.

Power Meter and Signal Strength Meter (S-Meter)

The power meter displays on a scale of 1–5 what proportion of the set's available power is actually getting out. If it says five when

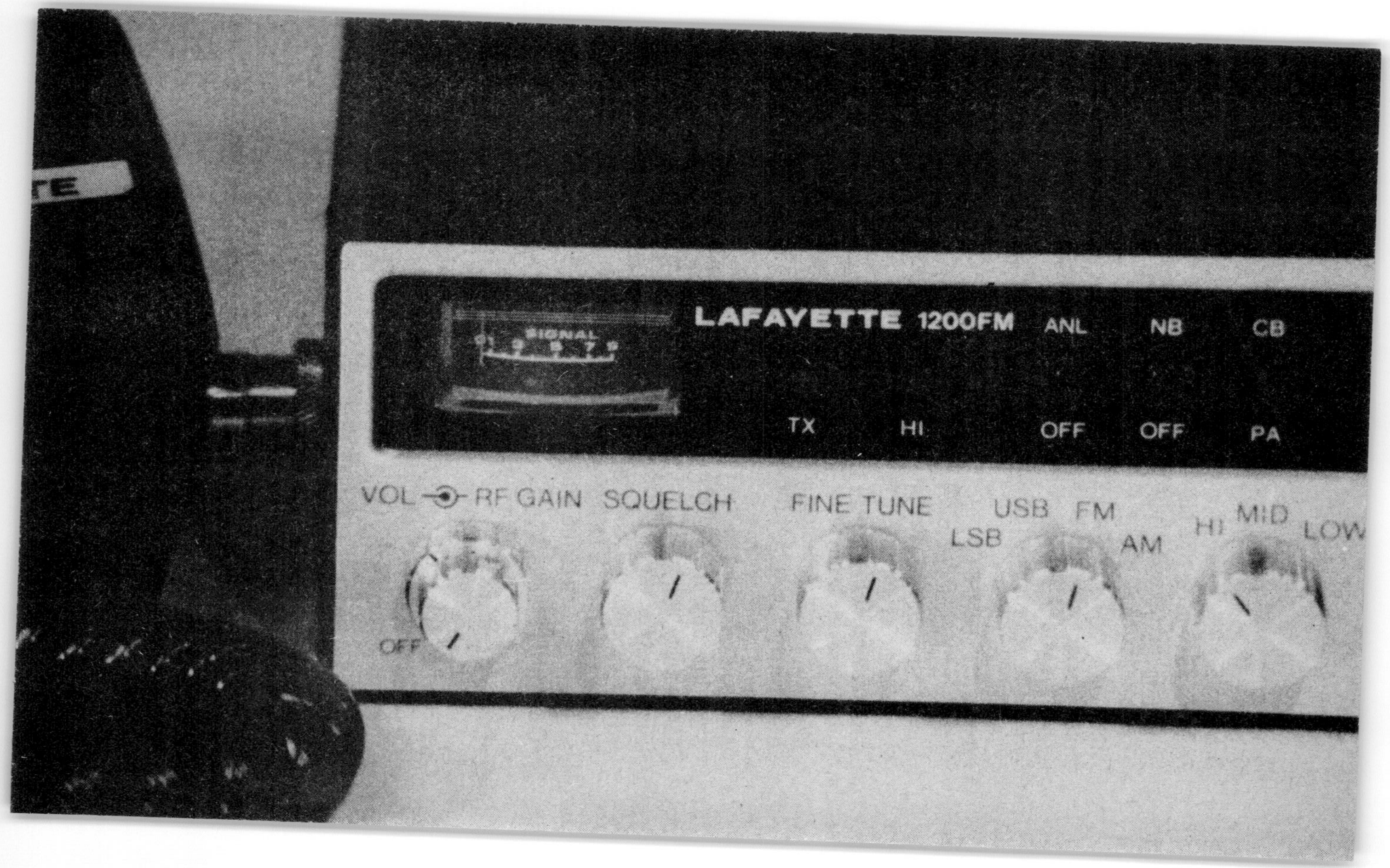
LAFAYETTE 1200FM
SIGNAL
ANL
NB
CB
TX
HI
OFF
OFF
PA
VOL RF GAIN
SQUELCH
FINE TUNE
LSB
USB
FM
AM
HI
MID
LOW
OFF

you press the PTT then you're doing fine and if it says nothing then the reason no one will answer you isn't because of something you've said. You haven't. The S-Meter shows on a scale of 1–9 the strength of incoming signals. If there's a car parked next to you transmitting like fury and only showing 1 or 2 on the S-Meter then one of you has a problem. With any luck it'll be him. The S-Meter reading is not related to the volume setting on the rig, so if it says 9 and you can't hear anything turn the set up or stop doing that and see a doctor.

Mode Selector
This allows you to choose between AM or Upper and Lower Sideband. Under the terms of MPT 1320 it is excessively illegal and shouldn't exist. Either you've just been done or you ought to be.

Delta Tune/Clarifier
This allows you to fine tune the receive side of the set in order to home in on transmissions which are slightly off frequency. Anyone who is should be told about it PDQ. In truth it is really only necessary when operating Sideband, which either means you've paid for something you won't need or that you obviously haven't learnt your lesson yet.

RF Gain
Yet another squelch-like refinement which, as its name implies, cuts down on RF gain in your receiver amp. Adjust it as you would squelch to cut down on background interference and distant stations and open it up wide when you need a break.

Mike Gain
This is the only thing which actually affects the way your transmission sounds. It operates a pre-amp connected to the microphone and enables you to hold the mike at a comfortable distance from your face and at the same time achieve a good output. You should decide how you normally hold the mike and get someone to give you a check while you adjust the gain to its optimum setting; you'll find there's a point at which your voice will begin to distort and other breakers will think you're talking rubbish. Of course they may be right . . .

61

CB/PA

Provided that a suitable external speaker is plugged into the correct socket this two-way switch allows you to alter your set from being a CB unit (while the switch is in CB mode) to being a PA system. In theory this handy option has very many uses of a socially acceptable nature, like crowd control at accidents and so on, but in practice it has far more entertainment value when used for anti-social purposes, like frightening the life out of innocent pedestrians and insulting fellow-motorists. As a general rule it should be treated with extreme caution unless you're the bloke who taught Bruce Lee everything he knew.

Roger Beep

On or off, this one, and preferably off. All it does is bleep at people when they're trying to speak to you. It's supposed to sound efficient and save time and effort, and while you might like it a great deal you could well discover that the rest of the world doesn't and has formed the silent majority without you. Still, maybe you're used to being ignored . . .

All of which wraps it up as far as basics are concerned. Naturally enough it's all going to get jolly sophisticated quite soon. Apart from anything else there seems to be a worldwide preoccupation with coloured lights, preferably of the flashing variety, just now, and almost certainly a large percentage of rigs will use coloured LED displays to indicate power output and signal strength as opposed to the traditional needle-in-a-window method. This appears to be no more or less accurate than any other way of going about the operation and is doubtless unlikely to be any more expensive, even though it looks better and more costly.

Many rigs also have a built-in SWR meter (which we'll deal with in the fullness of time) and these too are rapidly being replaced by little flashing numbers rather than a trad scale. Some of them even monitor their own SWR continuously, display it when asked and automatically switch themselves off if a fault develops which puts it above an acceptable level. They usually inform the operator by flashing a series of zeros or the letters SWR that they have taken this step on his behalf. Thoughtful, that.

Newer sets will also be likely to have an automatic squelch

control which can be set in the normal way and then pushed to go into auto mode. They shut the squelch right down, allowing you to listen to Radio One or whatever, until they receive a signal with a strength higher than your pre-set squelch threshold, at which point they open up to that level and let you monitor the transmission and make your own mind up whether you want to answer or not.

This is not very different to the tone-squelch system which has been around for ages. The same procedure applies, except that your squelch will only respond to a pre-determined tone. The trouble is that so many tones in the human voice correspond to the tones in use, so it isn't always that selective. And what with Roger Beeps as well . . .

Selective call has made an appearance as well. Rigs like this have a five-digit code number, which you choose for it, and when switched into selective mode will only respond to a similarly equipped set whose operator knows your number and has dialled it up. Not very different to the telephone, really. This type of rig will also scan for busy channels if you wish to make small talk, or scan for empty channels if you want a quiet chat. Some can be trained to keep a constant watch on channel 9 (the designated emergency channel) and will switch to it automatically if it gets busy. You can get rigs which will monitor any channel you choose, not only 9, or sets which will go to a chosen channel at a particular time, or just switch themselves on at a given time, all of which you may decide is likely to be useful to you in some way while driving. None of them can remember the name of that nice little French restaurant at Shepherd's Bush or even how to get there. They won't check your oil and water and wash the windscreen, either. Yet.

After all that you may be dizzy. No doubt you won't be the only one. But we still haven't looked at all the various gadgets you may wish to buy in addition to your rig. Some of these may be necessary, some not. Once again it's the only chance you'll get to make a choice so you'd best sit back and enjoy it.

Power Mike
This is dead handy if your set doesn't already have a mike gain control. A power mike has its own battery and pre-amp and may be adjusted in exactly the same way and for the same reasons as an

integral mike gain. A good power mike will almost always be an improvement over the one which came with the rig, if only because it will be of the crystal rather than dynamic type. You can buy one in any CB shop but you will almost certainly have to spend a while with a soldering iron connecting the leads to the plug; generally instructions enabling you to do this without causing explosions are included with the mike.

Speech Compressor

This is very similar in effect to a power mike but slightly superior. As the name implies it compresses the wave patterns made by your speech rather than actually squeezing you round the wind-pipe, and produces a clearer signal on the air. It helps you to modulate better in the correct and slang senses of the word. It won't actually make you sound like Paul Newman, but you probably didn't want to anyway. It is possible for this device to give you a little too much help, so like a power mike it is adjust-able, and often has a meter built in so that you can set it correctly without help and monitor its actions.

SWR Meter

If this isn't already built into the rig then you are going to need one. You only want to use it briefly so you could always borrow one, but since they're only about a fiver you may find it more convenient to buy your own and keep it around. Some of them are more complex than others, having dual or treble function, but once again that's only window-dressing above and beyond the call of necessity. You will have to use this device to tell you the relationship between your rig and your antenna before you begin transmitting, but it's dead simple. When you buy one make sure you also buy a patch lead as well, because you can't connect it up without one.

External Speaker

Most rigs will have a speaker built into the case. As a rule it won't be very large and once the rig is installed in, on or under the dash it won't be very loud either. An external speaker will cost very little, may be mounted in a much more practical and audible location and can be connected in seconds via the jackplug fitted to the lead. The manufacturers of your rig will very thoughtfully have provided a socket for this purpose and you will find it exceptionally beneficial if the socket and the jackplug are of roughly similar dimensions.

PA Speaker
Again this is easily fitted via a convenient jackplug, and once mounted under the bonnet of your car by means of two simple screws, will afford you hours of endless pleasure at pedestrian crossings, outside pubs late at night or when the driver in front is behaving like a berk. Just make sure he's a small berk is all . . .

Field Strength Meter
Now here's a useful little gadget. Not much bigger than a cigarette packet, but far cleverer. It buzzes at you when it picks up RF output from your antenna and doesn't when it doesn't. The best ones will measure the output for you, and tell you if you're actually putting anything out. Simply walking round your car or

A simple SWR meter. In fact it's so simple and easy that words are unnecessary. The controls on the fascia speak for themselves.

base antenna will tell you what sort of signal you're putting out and whether or not it's being masked. Unless you are related to Twizzle you will find the services of a friend/wife/granny a distinct advantage while operating it, because someone's got to push the transmit button.

Dummy Load

Transmitting without an antenna has two drawbacks. One, it's a pointless exercise, and, two, putting the antenna back in place afterwards is equally pointless since there won't be very many output transistors in working order left inside your rig. There are times when you might wish to test various functions of your rig without blotting out all the breakers in the area with a dead carrier. A dummy load, which once in place will perform a very creditable impersonation of an antenna as far as your transistors are concerned, means that you can do this. And if you're having trouble getting your SWR right and think that the fault may be with the rig itself you can confirm this by matching the rig to a dummy load. All it is, the dummy, is a 50 Ohm resistor, which is precisely what a perfect antenna should be.

Frequency Counter
This does precisely what you'd expect; it counts to several places of decimals your transmit frequency and tells you what it is. This information can be valuable for checking that your rig is perfectly tuned, but remember that there are 40 channels, so you'll either have to remember all 40 frequencies or look them up. Mostly this is useful as a base-station add-on or for performing static checks on your mobile installation. Its use is not recommended to motorists on the move, as they should be much too busy changing gear and generally looking where they're going.

Preamplifier
Connected between the antenna and the rig this will amplify the volume of all incoming signals, and while it may sound like a truly wonderful invention it is not, regrettably, without drawbacks. Like so many electric gadgets a pre-amp is basically stupid, and cannot tell the difference between things you want to hear and things you don't. In consequence of this it amplifies everything indiscriminately, including static, ignition noise and all kinds of interference. They are much more useful as an addition to a base installation, especially as many of them have a tuneable gain control which is not a feature of all base rigs.

Linear Amplifier
Does what a pre-amp does only going instead of coming. You already know these are illegal, so why are you interested?

Birdcalls
The minute the FCC made these horrid things illegal in America they turned up over here, where we have not yet had the sense to outlaw them. They produce a series of electronic screeches which are supposed to sound like birds but which actually resemble a Komodo dragon in distress rather more closely. The original justification for their production was that they attract attention and there's no doubt that this is true. But there are better ways . . .

When you buy your rig there is one choice which perhaps you should make before you even think of parting with the cash, and that is whether you wish to install it yourself or whether you're going to let someone else do it for you. If you go for the latter option then you will probably be better off buying it from a place which will then fit it free; there are any number of such enterprises all round the country. The better ones will very likely be members of CRISP (Car Radio Installation Specialists) and will display a sign to this effect. The absence of such a sign does not immediately indicate that you are on the premises of a cowboy outfit, but it's something to look for. If you're fitting your rig yourself then buy where you get the best deal. If it's a home base then you are at present very much on your own anyway, although there is no doubt that competent radio engineers will soon be offering an installation service for CB base units.

Meantime let's treat the two separately and look at cars first, since the principles are much the same. Cars will present you with more problems than your living-room or wherever, anyway, since they are both smaller and noisier.

From an electrical standpoint cars only come in two varieties; positive earth and negative earth. Fortunately CB rigs may be of both types also. They may even be adjustable. Before you do anything else make sure that your car and your rig share a common polarity. If they don't then your rig will not work. It will catch fire instead. This expensive disaster is to be avoided at all costs, so avoid it.

The first thing is to buy a unit which will actually fit somewhere within your car where it can be seen and operated. There's no point in having all the latest silicon-chip technology with flashing LED displays if it's bigger than the biggest empty space in the car. If it's going into an existing hole in the dash (like the one provided for music radio sets) make sure the set is the same size as the hole before you buy it and thus save yourself untold embarrassment and aggravation.

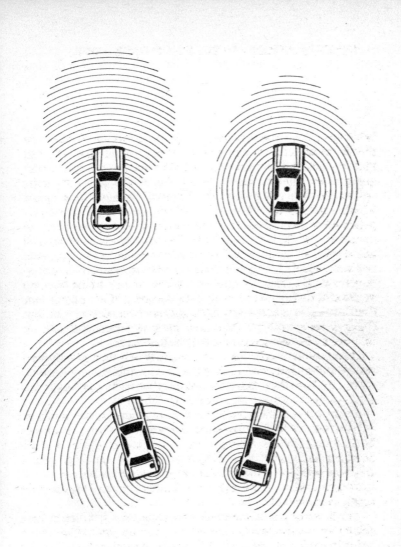

The difference in radiated signal caused by mounting your antenna in various different places on the car can easily be seen with the help of a good artist. In truth it won't make a great deal of practical difference to your enjoyment of CB and far more mundane considerations are likely to affect your choice of mounting site.

The precise location of your CB is more important than you may at first believe, and your decision in this matter should take several factors into account. To begin with you will need to be able to both see and reach it without affecting your ability to control the car and operate its vital functions successfully. This makes an in-dash or below-dash location as close to the driver's seat as possible the most desirable choice. Unfortunately the area behind car dashboards is usually either totally inaccessible or full up with Very Important Things Indeed. Do not damage or destroy any of these Things. You may not live long enough to regret it.

Be careful not to mount the rig itself anywhere near heater outlets; the transistors may not actually melt, but they won't throw a party in your honour either. And bear in mind, if you can, that even if you can afford the most whizzo, expensive rig ever made the world is full of people who can't but who would still like one. Yours. Aside from welding the unit to the dash (in which case they'll probably just nick the entire car) the best way to prevent theft is not to invite it in the first place. The only way to achieve this is to conceal your rig from view, either by mounting it in the glovebox (where you won't be able to operate it) or behind a false fascia plate on the dash (very clever, that) or putting it under the seat (where it will almost certainly be trodden on). Alternatively you could install it on a slide-mount, which allows you to remove it from the car in milli-seconds if you park it in the street or a public car park. Possibly the idea of wandering through the local shopping centre clutching eight pounds of useless junk every Saturday afternoon doesn't have much appeal. Lock it in the boot.

The next thing you should worry about at this stage is the antenna. As far as MPT 1320 is concerned you're rather short on options as far as type and performance are concerned, but your biggest problem at present is far more likely to be where you put it. In fact the location of the antenna can have some influence on the way your signal gets out. This is because the metal body of the car acts as a groundplane to the antenna. Consequently the strongest signal from the mast will be radiated in the direction of the longest line which can be drawn across the car and passing through the antenna. In this instance the pictures are probably worth about eight million words each.

It's obvious that the best location for your antenna will there-

Centre roof mounting. Although you can't see it this one's on a mag-mount base; you can probably spot the co-ax coming through the passenger window; this will always be a problem with semi-permanent fixings.

fore be right in the centre of the roof. Even to a dedicated breaker this may not be altogether acceptable. To begin with it's right out of the question for soft-top cars. Furthermore, making holes in the middle of the roof is neither the easiest thing in the world nor the driest. Thus mounting locations tend to be either on a wing top, just like a music radio aerial, or sometimes in the centre of the bootlid scuttle or bootlid itself. However, some of us might not be overkeen on drilling vast holes in the metalwork anywhere on the car at all, in which case there are several alternatives.

The easiest of these is the mag-mount. This is quite simply an antenna fixed to a magnet of dramatic proportions and capability. It will glue your antenna to any metal portion of the car you choose and keep it there for as long as you wish. At motorway speeds it appears unlikely that it can stay in place, but it does, and you will soon acquire faith. It allows you to fix the antenna to the roof of the car where it really ought to be without the necessity for excavating the bodywork, which is an entirely satisfactory state of affairs. It is also thiefproof and vandalproof, since removing it and placing it in the car is only the work of moments.

It is not the only antenna mount of its kind either. There are various means of fixing a mast to your car without making fixing holes, usually via a screw-up clamp. This is not as easy to remove as the mag-mount, but provides for mounting on bootlid, rain-gutter, bumper or other suitable protuberance; you may choose whichever suits you best. Like the mag-mount all of these semi-permanent arrangements have their drawbacks, only two of which will be likely to worry you unduly.

To begin with, a temporary installation is just that, and if it is your desire to avoid drilling the car for the antenna mount itself you must also appreciate that you will not be drilling holes in order to allow the feedline access into the vehicle. Therefore you will have to admit it via some other orifice. Nine times out of ten this means through a window or the edge of the door. Several tribulations will arise from this. You will never be able to close the door or window properly again, which might be fine in Southern California but won't be so funny in Arbroath on a cold December night. You will always run the risk of the cable being cut or damaged at the point of entry and if you wish to avoid the consequent damage to your rig which this will entail you will have to inspect the cable very carefully at frequent intervals. And as a

72

The truckers' favourite mounting – on the mirror arm. In many cases this type of fixing is done in pairs – one each side, fitted as a matched pair of co-phased antennas. Looks good and works very well indeed.

final bonus there will always be great loops of cable trailing all over the floor of the car for you and your passengers to trip over and/or break. You can (and should) tape this back with something at least as strong as the 200-mile-an-hour tape drag racers use, but even this is unsightly and could hardly be considered permanent.

More interesting is the question of earthing. For radio antennas a strong earth to the car body is essential for maximum performance. As far as mag-mounts, gutter-mounts and all other kinds of semi-permanent mounts are concerned this is almost impossible to achieve, which means you can expect some drop in performance compared to a properly fixed mast.

Mobile antennas. Although they're fitted with a conventional fixing, which can be angled to suit your needs, the wing-nut arrangement allows them to be fitted to a magnetic base or a gutter mounting.

74

If you decide to take the plunge and fix your antenna by drilling holes in the bodywork then the first rule is to do nothing until you have thought it all through very carefully indeed. Holes are easier to make than they are to fill in.

Choose the location and then, before you do anything else, make sure that there is only one thickness of metal at the site and the car isn't double-skinned at that place. Then make sure you have sufficient room beneath the hole for all the bits which hang down out of sight. This applies especially to telescopic or electrically-powered antennas, which may require up to 12 inches of free space beneath their mounting. Then check that the cable will reach the rig from the point you have chosen; that it is long enough physically and that there are no insurmountable obstacles between the rig and the antenna. Some bulkheads may be double-skinned or full of impenetrable things like door-hinges, seat-belt anchorages or other vital impedimenta. Only when you are absolutely sure that everything is going to work out according to plan should you get out the Black and Decker.

Think about all these things before you decide where you'd like to put your rig. Then, before you take any drastic steps, make sure that the installation will be possible. Can you screw it in place without breaking something important? Can you run a power line to it without the need to rend the vehicle into its component parts? Can you run the antenna feedline to it as well? Remember that co-ax cable is quite thick and that there is a reasonably substantial plug on the end of it which may require the creation of reasonably substantial holes in bulkheads and similar. If this is so make certain that there is nothing vital on the other side of the bulkhead and that you can reach the required place with the pointed end of a power drill. Cars are a lot stronger than you think. It's no bad idea to lay all the various bits of the installation out in the car and check each piece of the operation before you do anything drastic.

Check that the set has a firm mounting within easy sight and reach of the driver, that all feedlines reach easily without stretching, kinking or chafing and then screw it all together.

The set will need to be earthed, and it is possible that this can be achieved through the casing, provided that the fixing screws go straight into the metal of the car bodywork or chassis. If they don't or if you're not sure then you will need to run an earth lead

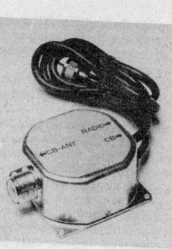

A splitter box. It allows you to use the same antenna for both CB and music radio. Remember that your radio will work a treat with a CB antenna but the reverse is not true at all. It's expensive. If you use one of these devices you'll find it's tuneable by means of a small screw. Trim this, using an SWR meter, before you adjust the antenna length. Be prepared for some drop in performance.

Another type of splitter box, which allows you to use up to three antennas in conjunction with one transceiver – one at a time, of course.

from the casing (there will be a fixing on the casing provided for this) to a metal part of the car. Nine times out of ten you will find that the metalwork behind the dash is already carrying a great number of earth leads from a great number of electrical devices and you could quite easily utilize one of these. To minimize the risk of interference it would be best if you made a fresh connection of your own.

Again to minimize the risk of interference you would be best connecting the power feed direct to the battery; this is sometimes clumsy and it does mean that you will have to remember to switch off the set when you leave the car. You may prefer to connect the power feed to an ignition-sensed terminal on the fusebox; make sure it doesn't also carry power to a potential source of interference like indicators or wipers.

Make certain also that there is a low-rated fuse (probably 2 Amps but the specification with the rig will tell you definitely) between the rig and the power source. If the power feed or the earth pass through holes in the bulkhead make sure that you use a rubber grommet to prevent them chafing. Don't leave the wires hanging loose; tape them back. Don't run them near areas of extreme heat, lest they melt, and try also to route them away from things like wipers, ignition coil and indicators.

The same rules apply if you are proposing to make any kind of mobile installation, either in a boat, a caravan or even an aircraft. Unless the caravan has its own internal power source you will need an extra tow-plug since there are unlikely to be any spare terminals on the one which handles the lighting. Don't be tempted to leave a loop of wire wrapped round the tow hitch and connected via a weatherproof domestic plug. It's obviously quicker, cheaper and easier in the short term but it's also inviting disaster. The best thing that can happen is you'll do the output transistors in and the worst is that you'll be toasting your dampers round the blaze and sleeping with that Al Fresco chappie under the starlight. Don't take chances.

Installing a rig in your house is easier, although it's not totally without pitfalls and if the thing that falls into the pit is your ancestral seat it can be more expensive than you bargained for.

The first thing is to decide whether you want a base unit for the sake of convenience or because you're interested in CB. If it's only convenient to have one, ask yourself, 'do I want it as well as a mobile unit or could I make do with a dual-purpose rig?'. If you only need it when you're at home then you could easily fit the mobile into your car on a slide mount and take it into the house when you're at home.

If you decide that this is the preferred method then you must make another couple of choices. You wouldn't, of course, be silly enough to plug your mobile unit into the household mains. Mobiles work on 12 volts DC. Houses work on 240 volts AC. There is a considerable difference and it is sufficient to convert your rig into a heap of red-hot rubbish. There are several ways round this.

You could bring the car battery in every time you bring the rig in, but this isn't the ideal solution. To begin with you have to be strong enough to lug the battery about without acquiring a double hernia and also fit enough to push-start the car every morning after the rig has used up all the power in it. You could buy another battery; you may even already have one. This is far more acceptable from a medical point of view but still not Utopia. Batteries, as you are probably well aware, smell. They also exude sulphuric acid as a gas, which is not recommended in bedsits or small rooms. Further, they drip said acid on the floor. It doesn't matter how tightly you screw the caps on, or how carefully you top it up, and it makes no difference that you never move them or spill the contents. They drip acid on the floor. It's like toast always falling butter side down. Sulphuric acid, as you are doubt-less well aware, does not mix well with floors, carpets, cork tiles, clothes, bed linen or small children. All these things will be generously coated with it in very short order no matter how meticulous you are.

Even supposing you're prepared to risk that, you will still find that the battery must be charged regularly. Just how regularly is something you will need to monitor closely, and this involves considerable amounts of intricacy with things called hydrometers. These are fiendishly clever devices whose sole purpose in life is to spread acid on your carpets. Anyone who is bereft of carpets, children and bed-linen etc., may still feel that having a battery about the house is not a problem. Using it could be.

All previous remarks about car electrics, CB rigs and polarity still apply, which means that you will need to make certain that your connections are the right way round if you wish to avoid

A power transformer. This neat unit turns mains AC voltage into low-power DC current of the type which will not cause conflagrations in mobile units doubling up for base use. So much neater than an old car battery, don't you think?

impromptu barbecues. And while anyone can get it right once, having to do it several times a day mutliplies the risk by an ever-increasing factor. Add in the necessity of connecting the battery to its charger at frequent intervals and the potential for disaster expands out of all proportion to the benefits. Actually that should be benefit, because as far as anyone can tell there is only one to be gained from using a battery; you'll still be on the air if there's a power cut.

It would be much simpler to buy a transformer (for less than the price of a battery) which converts mains voltage into 12 volts DC. This can be permanently connected to a mobile unit left in the house or simply fitted with two different sorts of plug, one on each power lead, thus preventing even an idiot from making an expensive foul-up on the occasions when the mobile rig is brought in from the car. Doubtless you will form your own opinions on this matter in due course.

On the other hand you may not find that a mobile unit, whether on a temporary or permanent basis, fulfils your needs for a base station, and therefore buy a purpose built unit. This solves a lot of problems, not least the question of power supply, since you can plug it straight into the nearest wall socket and stop worrying. The only thing to make certain of is that you use a three-pin plug and make sure that the earth to the socket you're using is par-ticularly good.

Now then. At this stage of the game your CB installation is ready to receive an antenna. You will probably have already chosen a site for this. With due consideration for the licensing requirements and the mandatory power reductions consequent upon mounting it more than 23 feet up in the air the results you get will improve in direct proportion to increases in its height above ground.

Rooftop mounting is a firm favourite; flat roofs are better than pointed ones unless you can fix it to the crest. If you use a chimney then make sure the antenna is taller than the brickwork and stacks, otherwise it will be masked by them. Keep it away from overhanging trees and as far away as possible from TV or domestic radio aerials already in place. Fix it as firmly as you can. Use masonry bolts, clamps, whatever will take the weight; if you're in doubt ask your hardware store for their advice; they will have a selection of fixing methods which were never dreamed of in anybody's philosophy but their own, never mind this Horatio chappie. It may seem expensive at the time, but apart from the inconvenience of having to climb on the roof to replace it once a week your neighbours probably won't take kindly to being trans-fixed by a six-foot metal pole every time there's a slight breeze.

It's quite likely that the roof of your house will be in excess of 23 feet high. If it isn't I'd like to borrow your ruler to measure my waistline. If it is you will have to seek another home for the antenna unless you wish to contravene the terms of your licence. Mounting it on the side wall of the house is not so difficult, but it does mean that your signal will be heavily masked for 50 per cent of the available compass points. Which half depends on which wall you choose. If you live at the seaside then you only have to decide between ships and motorists; anywhere else in the country and you've got problems. Look at a local map. Is there a big town

Complicated arrangement, this. Hardly what you could call a typical beam antenna, but don't worry. Under British regulations you won't be allowed it anyway.

vn you live in on one side
vay or main road in one
lorry drivers all the time?
? Decide what's important

o you like best and you'd
iile you could always stick
Choose a site as far as
ke sure that the pole isn't
s securely bedded in the
veight, but you still don't
e marrows, do you? Or,
sing guy wires to support
se three. If they are metal
et or so until they are at
wise they may degrade the
y think you're using them
know, against the law.
e rig you must use proper
ill suffice unless you want
in your garden. Without
ntenna and you will get
ybe up. In smoke. If the
more than about 100 feet
op in performance is the
stance to the passage of
ordinary co-ax begins to
mounts. You will have to
as lower resistance. Basic
getting less of something
d co-ax is no exception.
balances.
e way down to the base of
on't be tempted to take it
window. It may be easier
face it, under these rules
p it can get. Don't leave it
get mown twice a year or
ut you might. Bury it in a
you have to make joins in

A desk mike for use with a home base. The PTT bar is across the front of the base.

the cable don't mess about with plastic connecting blocks or insulating tape. Buy a pair of proper connectors and make strong, soldered joints. And if the joint is outside the house then cover it with some of that vile caulking stuff to make sure it's waterproof and then bind it with plastic tape.

If you do make a joint outside it would be the ideal place to install a lightning arrestor while you're at it. The aesthetic appeal of lightning diminishes in direct proportion to its proximity. The closer you get, in other words, the nastier it becomes. Close to it has a destructive capability surpassing even that of Frank Spencer, and will make large holes in brick walls with no apparent effort. Generally speaking it is not the sort of stuff you want knocking about the house on a Sunday afternoon, but unless you take precautions that is precisely where you will find it.

Lightning always strikes the highest point in an area. It also seeks the quickest and easiest path to earth, generally via some tall metallic object. Sound familiar? That's right. You've just built a thirty-foot lightning conductor in the back garden and led it straight into the house. Fit a lightning arrestor and lead the wire from it to a very good earth, preferably a steel rod sunk at least six feet into the ground. Run the wire between the arrestor and the earth-rod in a dead straight line. If you can manage it without getting silly make a few sharp 90-degree turns in the antenna feedline after the arrestor and before it comes into the house. And if there is a thunderstorm in your area it won't hurt to disconnect the feedline from the rig and sling it out of the window. Don't worry if you've got visitors who accuse you of being a yellowbelly. What would you rather be – a live chicken or finger-lickin' good? Lightning has a nasty habit of killing people.

Once the co-ax is in the house don't cut it to the exact length needed to reach the rig. Leave some spare cable coiled into tidy loops against the day when you finally get around to decorating and find that the rig would be a lot better off in that corner instead of this one.

Okay. Wherever you've put it – at home, in the car, in the office or in your personal bomb-shelter, your installation is complete and you're ready to switch on and start modulating.

Don't.

There is still one vital thing left to be done. We have already discovered that radio antennas work best if there is some corre-

84

spondence between wavelength in use and antenna size. This may have been an understatement. You are aware that transmitting without an antenna will have pyrotechnic results inside your rig. Transmitting with one which is the wong size bears similar penalties. If your mobile antenna has been vandalized or your base antenna gets eaten by the starlings don't even think about transmitting. Bin it and buy a new one. Promise, it'll be a lot cheaper in the long run.

Likewise do not transmit on a new or altered installation until you have made certain that the rig and antenna are matched as closely as can be arranged. What you must do is check, and adjust if necessary, this relationship. For this you will need time, patience, the little adjusting tool supplied with the antenna and an SWR meter. Also you will need to follow a few simple rules. The most important of these, and the most difficult, is simply this: trust me.

SWR is not a complicated thing at all. It only means Standing Wave Ratio and is a convenient way of measuring and describing how closely an antenna (which is necessarily mass-produced and therefore subject to all the usual vagaries of this process) matches the characteristics of the unit it is attached to. At one time, in the early days of pirate CB, a rig having an SWR below 1.5:1 bestowed respect and prestige on its owner roughly equivalent to his being in possession of the Cullinan Diamond. In fact it's not that tricky to attain and these days less and less people worry about it. Most Americans, who've been living with CB for more than 30 years, don't really care about it at all. Some don't even know what it is, but it doesn't seem to have affected their lives or the pleasure they get from CB. And even though it makes a difference to your rig a poor SWR is far more likely to be the result of bad installation or a damaged feedline than maladjustment of the antenna. The best SWR imaginable would be a perfect match – a relationship of 1:1. Income Tax will be abolished before you achieve this. In practice anything under 2:1 is acceptable and anything below 1.5:1 gets a mention in despatches.

The same principles apply to both mobile and base antennas, and the same means of checking and adjusting are required. A home base is easy enough, although for best results SWR should be adjusted once the installation is complete, which is fine if the antenna is on the roof or screwed to a wall, but not so clever if it's

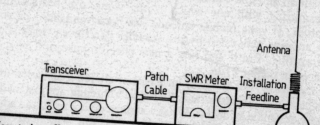

How to install an SWR meter. Dead easy. You couldn't get it wrong now, could you?

at the top of a flimsy pole 25 feet high. Doubtless you'll find a way . . .

So. Park your car in a bit of open space, away from overhanging trees and tall buildings, all of which may affect what you're about to do. Examine your SWR meter. You will see that it has a scale on the front, calibrated 1–9, a knob which looks like a volume control or similar, a two-way switch, marked 'forward' and 'reflected' (or just FWD and REF), and two screw connections, one marked CB and one marked ANT.

Switch off your rig and then disconnect the antenna feedline from it. Connect it to the screw connection on the meter marked ANT. Connect the little patch lead to the screw connection on the meter marked CB and then connect the other end of the patch to the rig. The meter is now installed in the antenna feedline between the rig and the antenna. Trust me.

You would do well at this point to close all the doors on the car as well as the bonnet and boot. Metal objects of this size in the vicinity of the antenna may have a slight but measurable effect on SWR, so unless you propose to drive around with them open all the time you ought to close them.

Switch on the rig. Provided that there are no loud bangs nor any sign of smoke it is safe to proceed to the next step.

Since each channel represents a marginally different frequency it follows that the correct-length antenna for each channel will also be slightly different to all the others. Thus if you do, by some miracle, achieve 1:1 on channel 1 it will be somewhat worse on

Combined power, SWR and modulation meter. Now you can see exactly what's happening to absolutely everything. This is for base use only, as you can see. It's too big for a car and you haven't got time to watch it all anyway.

channel 40. Since it is impossible for your antenna to precisely please all of the channels all of the time you have two choices. Either switch to channel 20, which will give you a good average across the band, or switch to the one you use most – 14 if you live in London, 19 if you're a trucker, 9 if you've never had a no-claims bonus in your life.

Switch the meter into 'forward' mode and thumb the PTT bar. Careful observation will reveal that the needle on the meter has moved and is now giving a reading of some sort. Don't panic, it's meant to do that. Ignore it, whatever it is, and twiddle the volume control lookalike until the meter gives its maximum reading. Then switch it into 'reflected' mode. It will give a different reading now, and this time it is important because it's telling you what your SWR is. If the figure is below 2:1 everything is fine. Switch off the rig, disconnect the meter, reconnect the antenna and go home. If it isn't, or if you have the sort of nimble fingers which enable you to construct delicate, intricate things like plastic models and time bombs, and you feel you could improve on it then it's time to get busy.

All antennas have provision for some adjustment in their length. This will either take the form of a screw-threaded tip with a locking nut or the whole mast will be held in the base by a locking screw and may be moved bodily up or down in its socket. If your SWR is too high then you will need to utilize this adjustment facility in order to make the mast longer or shorter as necessary. At this stage, however, you don't know which of the two alternatives is the right one. The easiest way to find out is put your hand close to the mast while you're thumbing the transmit switch and watch to see if the meter reading goes up or down as a result. Just how you're going to manage this with the doors closed is your problem. Quite likely Max Bygraves was right; you need friends, or one at least.

When you do achieve it all will be revealed. If the SWR goes up then the mast needs to be shorter and if it goes down then it needs to be longer.

Alternatively you could just alter the length of the mast up or down and then read the SWR again. If it's lower you were going the right way and if it's higher you were going the wrong way. Be warned that tiny alterations to the length of the antenna will have much larger effects on SWR, so only adjust it in little steps. Remember also that there is an ideal point and if you go beyond that point the SWR will start to climb again, which could fool you completely. You wouldn't be the first. Some whip antennas may be too long even when they are seated at the bottom of their socket and these will need to be trimmed with sharp wirecutters. But only minutely. Once you've cut a bit off you'll have big trouble trying to stick it back on, even with Superglue, so be sure before you start cutting.

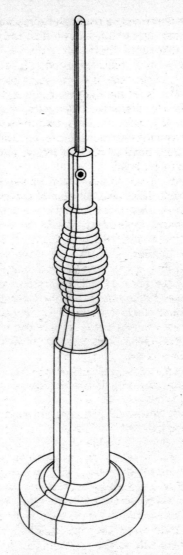

The bottom end of a typical base-loaded antenna. You can clearly see the Allen-head locking nut which allows you to release the tension and raise or lower the whip in its socket in order to adjust your SWR.

Once the reading drops below 2:1 you can pack it in unless you're dead stubborn or have these terribly nimble fingers, etc. Otherwise you could be at it for days, in which case I admire your stamina. What do they feed you on?

When you're sure you've had enough switch off, disconnect the meter, remake the antenna connections and you're ready for action. Leap into the car, start the engine, switch on the rig, stop the engine.

What was that noise?

At this stage of the game it's hard to say, but whatever it was it shouldn't worry you. The receive side of your rig is picking up all sorts of interference from the engine, and getting rid of it can be a time-consuming and complex business. And you thought SWR was difficult . . .

6 TROUBLESHOOTING – INTERFERENCE, SUPPRESSION, MAINTENANCE

We already know that the air is full of all kinds of radio signals which a CB radio receiver is quite capable of picking up, even if you don't specially want to listen to them. These usually appear through the speaker as an odd selection of crackles, whistles and pops. Equipment manufacturers have already provided you with a variety of refinements designed to reduce this row as much as possible. What they can't really do much about is the huge and overwhelming amount of electrical garbage which indicates the presence of an active petrol engine. In fact there are few places less suited to the installation of a radio transceiver than inside a motor car.

To a large extent this problem will be more aggravating to naughty people operating illegal AM rigs than it will to anyone on the legal FM service. This is because FM rigs will not make all the row through the speaker which AM users have come to know and love. As a rule they will remain silent, but their performance will suffer. However unpleasant the AM sound-effects were they did at least announce to the operator that a problem existed; if FM rigs are simply going to sulk then life could get very tricky indeed.

If you've got an AM set then you will know straightaway if you have a problem with engine noise. If you're on FM then you'll have to take the trouble to find out. All you've got to do is find a nice quiet, open space, well away from any potential source of interference. Turn the engine off, turn the ANL or NB switches on the rig off and turn the squelch right down and then tune in to a faint station on any channel – it doesn't matter which. Then start the car engine. If there is no change in the strength of the incoming signal then you can relax, because you haven't got any problems. If, on the other hand, the signal gets weaker or dis-appears then you know you've got some interference from some-thing on the car.

In a lot of cases the bulk of the interference arises from poor installation or damage to the feedline and can be cured quite simply. Earthing the rig and the antenna properly is most important and if this is not done then you will never cure the interference. Start with the antenna. Make sure that it is firmly fitted to the vehicle and that it has a strong, clean earth connection to the bodywork metal – not to the paint. Then check the feedline for cracks, splits, cuts or breakages. If it is compressed or kinked, free it off. Make sure then that the connection to the rig is sound, and that the transceiver itself is firmly earthed to bare metal. With any luck the situation should have improved. If it hasn't you will have to discover which bits of the car are making the racket, and this is simply done by a process of random selectivity, or trial and error, as it's better known.

Switch off everything electrical in the car except the rig. Now start the engine. AM sets will make noises, FM sets will go very quiet. Fortunately there aren't many electrical things connected to the engine which cause interference. The most likely source is the high-voltage spark generated by the ignition system and present at all of the plugs and inside the distributor. Ignition coils also operate at high voltages so that will need silencing as well. The typical symptom of this will be a vicious crackling related in intensity and ferocity to the engine speed.

First make sure that your spark plugs are fitted with all-enclosed plastic suppressor caps rather than simple metal clips. If they're not you will have to buy some. Then check the ignition leads. If they're copper-cored they should each have an in-line suppressor. If they haven't got these either get some or replace them entirely with the suppressed carbon-fibre leads. If you already have carbon leads then look at them closely. They are very susceptible to damage caused by heat and oil, and you may find that the insulation is breaking down. If they are cracked and/or oil soaked, replace them. In all probability this will have beneficial effects on the performance of your engine as well, which is just another of the little-known side-effects of CB.

If the problem persists then you should fit a suppressor to the coil. Any garage, motor accessory shop or CB shop will be able to sell you one for a few pennies. Suppressors are really only capacitors, which electricity, not being overburdened with intelligence, thinks is a discharged battery and rushes into. Once in, it stays, rather than galloping about into places where it is unwanted, like your rig. Bolt the body of the suppressor to a good earth – it's

92

simple to use the coil mounting bolts for this – and connect the wire (suppressors only have the one) to the terminal of the coil which carries live feed from the ignition switch. This terminal should be marked SW, which makes it easy to find. It may be that your coil terminals are only marked plus or minus, in which case look to see whether your battery is positive or negative earth. Whatever it is the live feed to the coil will be the other one. So in negative earth cars the live feed to the coil will be connected to the positive terminal. Easy. Pity it hasn't shut the interference up.

FM users will still only know it exists if they take the trouble to find out. AM users will be able to hear it, and the chances are it will, by now, be a whining noise. The generator. This too can be suppressed. Alternator suppressors should be connected to the live terminal once again, and that's easy because you'll only have a choice of positive or negative to make, and you already know the answer. Dynamos, if that's what you've got, only have two terminals. One big one and one little one. They're both on the back of the unit, the big one near the centre and the little one towards the outside edge. Do not, under any circumstances, connect the suppressor to the little terminal unless you really can't face another day. Fit it to the big one.

The engine should now run in total silence without disturbing your CB. If it doesn't, don't panic. Check the main battery earth lead and connections at each end. Clean and/or replace as necessary. Check the earth strap between the engine and the chassis. Clean or replace this too. Check that the power lines to the rig and the feedline to the antenna do not pass any closer to generator, coil, distributor, spark plugs or any high-tension wiring than they have to. Re-route them if they do, even if it means running wires all round the edge of the engine compartment. If the rig isn't connected direct to the battery already it may have to be from now on. You may also need to fit an in-line suppressor between the rig and the battery in order to eliminate conducted noise. If you still have problems you can buy special shielded distributor caps and spark plug covers which ought to fix you up properly. It may also be worth fitting a braided copper earth cable between the bonnet (fix it to the hinge) and the vehicle chassis to increase the shielding effect of the bonnet. A similar arrangement between the antenna fixing (below the bodywork, of course) and the car body may also help.

It's quite likely that owners of glassfibre cars will have to take all of these measures, since the glassfibre does not shield the antenna or the rig from the radiated interference generated by the engine. Mounting the antenna as far from the engine as possible is helpful, but many people with cars like this end up with all the suppression available, plus a metal sheet glued to the underside of the bonnet and earthed direct to the chassis, and even, in extreme cases, with their rigs mounted inside a sheet metal box which is also earthed to the chassis. And even overkill like that doesn't always work.

Hopefully it has, though, and the engine will no longer affect your rig. It's not quite over yet, though, as there are several bits of the vehicle's electrical establishment which may do so quite independently of the engine. To check this, switch the engine off and the rig on and then operate every one of the electrical devices fitted to the car one at a time. The most likely things to interfere are wiper motors, heaters, air-conditioning pumps (you should be that wealthy), indicators and perhaps even the relays which operate the headlamps. If you're really unlucky even the fuel gauge can be a nuisance. It sounds like a clock ticking and only starts when the needle on the gauge reaches the level appropriate to the amount of petrol in the tank. Other dash instruments might also affect the CB and these tend to make a hissing sound.

All of these things can be suppressed quite easily by connecting a suppressor to their live terminals. Wiper and heater motors are best suppressed on the unit itself, indicators should be suppressed on the flasher relay which you will have to track down by listening to it tick while the indicators are operating. It's probably somewhere under the dash. Probably . . . The fuel gauge can be handled either at the dash or on the sender unit while other instruments should be treated individually. Be very wary, incidentally, of electronic tachometers. They are sensitive little things, and not fond of being messed about with.

This process, if you've been forced to go through it from start to finish, will probably have taken days rather than hours. Possibly you never want to look a suppressor or a spark plug in the eyes again. Beyond any doubt you are bleeding from at least one part of your body, maybe several. Regrettably there is, as yet, no alternative method of achieving the same result, which is silly when you think of all the expertise needed to produce both car and rig. Still, they haven't cured the common cold either, despite the fact that heart transplants were mastered years go. It's always the glamorous things which get the most attention while the boring dull stuff tends to get ignored, although half the time life would be a lot nicer if it were the other way around. Which is a rather contrived way of mentioning that now your rig is installed and working, and free from interference, it would be nice to keep it that way, would it not? Even if it is boring?

Once a month if you can manage it, but definitely no less than every three months, have a quick check round the installation to make sure it's working properly. Check and clean all the earths – antenna, rig, battery and engine. Replace the leads if they're beginning to look tatty. Make sure the antenna feedline is un-damaged and still securely tied or taped out of danger. Make sure that the power feed to the rig is secure at each connection, that it too is undamaged and in no danger of shorting out, and that it is also securely held in place and under no threat of being pulled loose. Check that the rig is still securely fixed in place and have a look to make sure that all the suppressors you fitted so carefully and at such great personal risk are still all in place and properly connected. You won't hear it if they're not, remember, and you may not be aware of a gradual decrease in the efficiency of your rig.

Also have a look at the antenna itself. If it's the telescopic kind it benefits greatly from a wipe down with a mildly astringent spirit. If it's damaged in any way then replace it before it's too late to save your output transistors.

Remember also that a rig with poor SWR, whether caused by a badly adjusted or damaged antenna or a damaged feedline, even if it doesn't burn itself out, will drive the neighbours mad since it is much more likely to be putting out TVI. You can check this by parking next to your own TV set and keying the mike. If the picture goes fuzzy, the colour fades or your sound overrides the broadcast sound then you're causing TVI.

FM sets are less likely to do it than AM sets, but in either case the problem actually lies with the TV, which is incapable of rejecting signals outside its own receive frequency if they are sufficiently strong. You can fit a low-pass filter between your antenna and your rig which will minimize the chances of this happening. If your TV is particularly prone to TVI, or if the local

96

breakers club meet next door you can also get a high-pass filter to fit between the TV and the aerial, which should eliminate the problem.

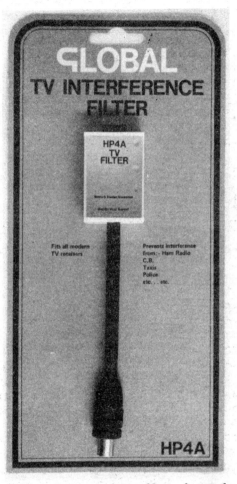

The other end of the story; an interference filter to fit just about any TV receiver which will prevent nearby or badly-adjusted CB units from blotting out Blankety Blank.

7 GETTING ON THE AIR – CODES, JARGON, PHRASES – RIGSPEAK – THE ART OF MODULATING

Up to now it's all been a bit new to you, right? All this business with SWR meters and TVI filters and whatever? And now you've arrived at the bit you've been waiting for. The easy part. Actually going out there and getting on the air, modulating with the other breakers.

Go on then. No, really, go on and try it. Don't worry about me, I'll wait here. See you in a minute or two . . .

That was quick. Didn't like it much, huh? Didn't say much either, huh? Never heard of mike fright before, have you?

It is, believe it or not, much more difficult than you think to just pick up a microphone and chat to a complete stranger, a complete *expert* stranger, via a medium you've never used before and in a private jargon you don't really understand. Which is why mike fright is so common. Mike fright is characterized by a total inability to speak or move coupled with an urgent desire for a cigarette/a drink/your Mum. You'll get over it.

The best thing to do is pretend your rig doesn't have a microphone attached to it. Just listen in – copy the mail as the breakers call it – to what's happening on channel. To start with it seems unbelievable that the world is full of people who speak in numbers half the time and riddles the rest of it, and who are continually asking complete strangers if they've got their ears on. But if you listen long enough it will start to make sense.

Sooner or later you'll hear someone asking for a 10–36 and when he gets six falsetto replies all about the third stroke you'll twig immediately that he wanted to know the correct time. If you're the observant type you will also realize that what he got instead was a wind-up. He still doesn't know what the time is, but at least he's laughing now.

A lot of the most common conversations will contain numbers; 10–1, give me a nine, what's your twenty, that's a four, one-nine for a copy. This is mostly because numbers, when spoken as individual digits (say channel one-nine, not channel nineteen) can be used to convey a great deal of information very swiftly with

98

much less chance of being misunderstood than long sentences. As long as everybody knows what the numbers mean. In the case of CB the numbers are loosely based on the Ten-code, which is a numerical form of verbal shorthand used by law enforcement agencies in America. Maybe you remember the TV series *Highway Patrol*. If you do then you will almost certainly recall Broderick Crawford growling 'ten-four' into his microphone in the same gruffly competent way that Gary Cooper used to say 'Yup'. By coincidence, ten-four and yup mean precisely the same thing. American policemen speak in numbers to each other all the time; they have number codes for every eventuality. On CB breakers speak to each other in numbers a lot, although not all the time. As a rule the sequence of number-exchanges tends to happen mostly at the beginning and end of conversations when time is important.

You will notice, if you tune to your local calling channel (if you don't know what it is already then find the noisiest – it'll be that one), that a great number of conversations or acquaintances are struck up and only last a matter of seconds, ending in a flourish of numbers. The reason is that a calling channel is just that; it's a place to meet people. It's a place which can only be occupied by two people though, which means that everybody else, while they may be able to listen to what's happening, cannot join in or use the calling channel for its proper purpose as long as it is being monopolized by one conversation. Therefore it is polite, having established contact with a breaker with whom you wish to speak, to clear the calling channel as quickly as possible, and move off to a vacant spot on the dial. After all, discounting the national emergency channel (9) and the calling channel itself, whatever it happens to be in your area, there are still 38 left to choose from. If you were to follow a couple of breakers when they leave the calling channel, and earwig at their chat on a different channel for a while you would discover that their conversation is almost normal and easily comprehensible. It is only because the calling channel is so busy that the forms of shorthand are necessary.

So listen to the chat for a few days. The numbers will soon make sense to you, and you will probably work out for yourself that the ten-code is the only time at which numbers will be spoken normally. You will hear 'break one-nine' or 'let's go to three-five', but always 'ten-twenty' or 'ten-thirty-six'.

You may also hear the odd bit of the Q-code creeping in;

certainly you will hear QSL, but perhaps other bits from time to time. The Q-code is mainly used by Ham operators working internationally and is much liked among Sidebanders shooting skip, but is unlikely to survive much longer on normal CB, so you needn't really be concerned with it.

When the moment you feel brave enough to speak arrives you'll find that it may be easier to answer another breaker looking for some gossip rather than initiate your own conversation. At least if you answer him you'll already know if you like the sound of his voice, and if he's a local then you may have heard him on channel before, and be reasonably happy to talk with him. Probably the most important thing to remember (and the easiest to forget) is that the chap at the other end of the conversation can't see you. In periods of silence he won't know if you've died, switched the set off, passed out of range or have simply finished speaking and are waiting for a reply. Therefore you must tell him. In cases of sudden demise, of course, you are excused this responsibility.

There are accepted forms of handing the initiative in a conversation over to the other party on CB, which may take the form of an invitation – come on, come back, bring it back – or of an interrogative – 'do you copy?' 'Copy?' 'Is that a four?' Whether you know what they mean or not it will be quite clear when you hear them spoken that you are required to make some form of reply.

Likewise the invitation to change channels will be brief and perhaps even obscure – 'let's go to', or 'let's take it to', or even 'let's take it up/down to', followed by a number – but all you need to remember is the number. If you get there and you've lost your contact go back to the calling channel and try to locate him again.

It's okay to break into the calling channel looking for a contact, it's okay to ask for a quick time-check on the calling channel, but discussing the weather or your health is not on. Do all of that anywhere else, except channel nine.

And when the time comes to close the conversation you'll find that there are a great variety of ways in which this can be done. None of them are anything like the polite, formal phrases you learnt in your Swiss finishing school. Basically they are divided into two types – finishing a conversation but staying on the air, or finishing a conversation and switching the set off. People doing the former will very likely say something like 'ten-ten till we do it again' and then claim to be 'down and on the side'. Don't worry, this is fairly normal and quite acceptable. People closing down may say any number of enormously complex things, but these will generally include remarks like 'breaker-break' or 'I'm down and gone' delivered with unmistakable finality. Prior to this they may wish you any combination of threes, eights, seventy-threes, eighty-eights, high numbers, lucky numbers or golden numbers, all of which are terribly friendly and enormously polite. The only one to watch out for is the eighty-eights, which means love and kisses. The number of occasions on which you hear this directed at you will depend entirely upon the gender and/or inclinations of the people you speak to. You have been warned.

After you've had a go at this you will doubtless want to start your own conversations, and by now you shouldn't need much help. You will already know that there are any number of ways to make a break on channel, all of which are acceptable and permissible. If the channel is reasonably free you can be leisurely about it, especially if you're just looking for a natter – 'any ears on channel', 'breakers any takers', that sort of thing – but if it's busy you'd better be quicker and know what or who you want – 'one-nine for a ten-thirty-six', 'one-nine for a southbound'. If it's very crowded the polite thing to do is quickly announce your presence during a handover – 'breaker on the side' – and wait to be invited into the conversation – 'go breaker'.

And now all of a sudden there you are having a wonderful time. One of the nicest things about CB is that it destroys a vast number of our social customs and barriers. When you speak on channel you will have access to a huge pool of new friends. You may know or have eyeballed one or two of them, but in the main they'll just be voices in the ether. They could be tinkers, tailors, soldiers, even spies. One of them may even be Ronald Biggs, but you'll never know. They won't even have names to identify them, since it is a long accepted practice to adopt a handle for use on the air. Some of the time the handle will reveal more about a person than his or her real name – Mean Machine was never the slowest driver in town, Mickey Mouse was a happy fellow, Sugar Plum was a pain in the neck and I never met Big Willie, but I know several people who'd like to.

A handle doesn't have to be a character-related nickname, though. It may be chosen strictly for laughs or based on your own

101

initials expressed via the phonetic alphabet – Delta Tango, November Delta – or be your favourite film or TV personality. Can't help thinking that anyone using Kojak as a handle is going to be shinier on top than the rest of us, though.

Choosing a handle for yourself can be quite difficult. Most of the best ones have been thought of already, you see. And asking friends for help is of no practical value at all; they've all got something they'd like to call you, but would you have the front to use it as a handle? No, this is one time when you're on your own. On a practical level remember that words which contain a lot of sibilants – which start or finish with the letter s or with a soft c – are easily misunderstood on the air. This doesn't mean that you shouldn't choose a handle incorporating them, especially if it's significant to you in any way, but you should be prepared to repeat yourself several times and still be misunderstood.

And after all that it would do you no harm at all to study and digest the Home Office Code of Practice, issued in October 1981 and compiled in conjunction with various interested CB organizations. If nothing else, it's brief and to the point and easily comprehensible.

The CB Code of Practice

Read your licence – it tells you what you can and cannot do. The conditions have deliberately been made simple with few restrictions. It is up to you to develop this service as you wish for the benefit of all. This means having consideration for one another and recognizing that no one has preferential rights at any time or place or on any channel. NATCOLCIBAR, the Parliamentary CB Working Party, and representatives of industry have in consultation with the Home Office prepared this simple code of practice. If you work to it you will help the system to help you.

1 Listen before you transmit. Listen with the Squelch control turned fully down (and Tone Squelch turned off if you have Selective Call facilities) for several seconds, to ensure you will not be transmitting on top of an existing conversation.
2 Keep conversations short when the channels are busy, so that everyone has a fair share.
3 Keep each transmission short and listen often for a reply – or

you may find that the station you were talking to has moved out of range or that reception has changed for other reasons.

4 Always leave a short pause before replying so that other stations may join the conversation.

5 CB slang isn't necessary – plain language is just as effective.

6 Be patient with newcomers and help them.

7 At all times and on all channels give priority to calls for help.

8 Leave channel 9 clear for emergencies. If you have to use it (for instance to contact a volunteer monitor service) get clear of it as soon as you can.

9 If there is no answer on channel 9, then call for help on either channel 14 or 19 where you are likely to get an answer.

10 If you hear a call for help, wait. If no regular volunteer monitor answers, then offer help if you can.

11 There is no official organization for monitoring CB and no guarantee that you will always be in reach of a volunteer monitor. CB is not a substitute for the 999 service.

12 Respect the following conventions – Channel 9: Only for emergencies and assistance. Channel 14: The calling channel. Once you have established a contact, move to another channel to hold your conversation. Channel 19: For conversations among travellers on main roads. (Remember, if you are travelling in the same direction as the station you are talking to, not to hog this channel for a *long* conversation.) Give priority to the use of this channel by long distance drivers to whom it can be an important part of their way of life. Other: You may find that particular groups in particular areas also have other preferred channels for particular purposes.

13 Use commonsense when using CB and do not transmit when it could be risky to do so. For example, don't transmit: (a) when fuel or any other explosive substance is in the open – e.g. at filling stations, when petrol or gas tankers are loading or unloading, on oil rigs, or at quarries; (b) when holding a microphone which may interfere with your ability to drive safely; (c) with the antenna less than 6 inches from your face.

14 Interference can be caused by any form of radio transmission. Avoid the risks. Put your antenna as far away as possible from others, and remember that you are not allowed to use power amplifiers. In the unlikely event that your CB causes interference, co-operate in seeking a cure using the suggestions from a good CB handbook. Moving the set or antenna a few feet may cure the problem.

103

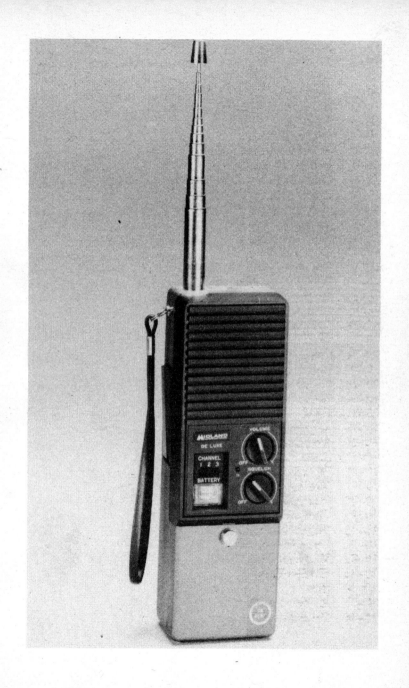

10–37	(Investigate) suspicious vehicle
10–38	Stopping suspicious vehicle
10–39	Urgent – use light, siren
10–40	Silent run – no light, siren
10–41	Beginning tour of duty
10–42	Ending tour of duty
10–43	Information
10–44	Request permission to leave patrol . . . for . . .
10–45	Animal carcass in . . . lane at . . .
10–46	Assist motorist
10–47	Emergency road repairs needed
10–48	Traffic standard needs repairs
10–49	Traffic light out at . . .
10–50	Accident
10–51	Wrecker needed
10–52	Ambulance needed
10–53	Road blocked at . . .
10–54	Livestock on highway
10–55	Intoxicated driver
10–56	Intoxicated pedestrian
10–57	Hit and run
10–58	Direct traffic
10–59	Convoy or escort
10–60	Squad in vicinity
10–61	Personnel in area
10–62	Reply to message
10–63	Prepare make written copy
10–64	Message for local delivery
10–65	Net message assignment
10–66	Message cancellation
10–67	Clear for net message
10–68	Dispatch information
10–69	Message received
10–70	Fire alarm
10–71	Advise nature of fire
10–72	Report progress on fire
10–73	Smoke report
10–74	Negative
10–75	In contact with
10–76	En route

10–77	ETA (Estimated Time of Arrival)
10–78	Need assistance
10–79	Notify coroner
10–80	Chase in progress
10–81	Breathalyzer report
10–82	Reserve lodging
10–83	Supervise school crossing at . . .
10–84	If meeting . . . advise time
10–85	Delayed due to . . .
10–86	Office/operator on duty
10–87	Pick up/distribute cheques
10–88	Advise present telephone number of . . .
10–89	Bomb threat
10–90	Bank alarm at . . .
10–91	Pick up prisoner/subject
10–92	Improperly parked vehicle
10–93	Blockade
10–94	Drag racing
10–95	Prisoner/subject in custody
10–96	Mental subject
10–97	Check (test) signal
10–98	Prison/jail break
10–99	Records indicate wanted or stolen

The 10 codes used on the CB band in America:

10–1	Receiving poorly
10–2	Receiving well
10–3	Stop transmitting
10–4	Acknowledged, message received
10–5	Relay message
10–6	Busy, stand by
10–7	Out of service
10–8	In service
10–9	Repeat message
10–10	Transmission completed
10–11	Talking too fast
10–12	Visitors present
10–13	Advise weather and/or road conditions

10–16	Make a pickup at . . .
10–17	Urgent business
10–18	Anything for me?
10–19	Nothing for you
10–20	Location
10–21	Call on landline
10–22	Report in person to . . .
10–23	Stand by
10–24	Completed last assignment
10–25	Can you contact . . .
10–26	Disregard last information
10–27	I am moving to channel . . .
10–28	Identify your station
10–29	Time is up for contact also I am leaving this location
10–30	Does not conform to FCC rules
10–31	Crime in progress
10–32	Radio Check
10–33	Emergency traffic at this station
10–34	Trouble at this station, help needed urgently
10–35	Confidential information
10–36	Correct time
10–37	Wrecker needed at . . .
10–38	Ambulance needed at . . .
10–39	Your message delivered
10–41	Please change to channel . . .
10–42	Traffic accident at . . .
10–43	Traffic holdup at . . .
10–44	I have a message for . . .
10–45	All units please report
10–46	Assist motorist
10–50	Break channel also traffic accident at . . .
10–51	Wrecker needed at . . .
10–52	Ambulance needed at . . .
10–53	Road blocked at . . .
10–55	Drunk driver
10–59	Convoy or escort
10–60	What is next message number?
10–62	Unable to copy, use landline
10–63	Network directed to . . . or prepare to make written copy
10–64	Network clear or network directed to . . .

10–65	Awaiting next message or network clear
10–66	Cancel message
10–67	All units comply
10–68	Say again
10–69	Message received
10–70	Fire at . . .
10–71	Proceed with transmission in sequence
10–73	Speed trap at . . .
10–74	Negative
10–75	You are causing interference
10–77	Negative contact or ETA
10–81	Reserve hotel room for . . .
10–82	Reserve hotel room for . . .
10–84	My telephone number is . . .
10–85	My address is . . .
10–88	Advise telephone number of . . .
10–89	Radio repairman needed at . . .
10–90	I have TVI
10–91	Talk closer to your microphone
10–92	Your transmitter is out of adjustment
10–93	Check my frequency on this channel
10–94	Please give me a long count
10–95	Please transmit a 5-second carrier
10–97	Check test signal
10–99	Mission completed
10–100	Personal reasons, rest-room stop
10–200	Police needed at . . .
73's	Best wishes, regards
88's	Love and kisses

Q code signals used by Ham operators internationally and also by Sidebanders to an extent:

QRA	What is the name of your station?
QRB	How far are you from my station?
QRC	Will you tell me my exact frequency?
QRD	Where are you going and where are you from?
QRE	What is your estimated time of arrival at . . .?
QRF	Are you returning to . . .?

QRH	Does my frequency vary?
QRK	What is the readability of my signals?
QRL	Are you busy?
QRM	Are you suffering from interference?
QRN	Natural interference – static
QRO	Increase transmitting power
QRP	Decrease transmitting power
QRQ	Transmit faster
QRS	Transmit more slowly
QRT	Stop transmitting
QRU	Anything for me?
QRV	Are you ready?
QRW	Shall I inform . . . that you are calling him on channel . . .?
QRX	When will you call me again?
QRY	When is my turn?
QRZ	Who is calling me?
QSA	What is the strength of my signals?
QSB	Are my signals fading?
QSL	Ackowledge receipt
QSM	Repeat last message
QSN	Did you hear me on channel . . .?
QSO	Communicate with . . .
QSP	I will relay
QSW	Do you wish to transmit on this channel?
QSX	Listen to . . . on channel . . .
QSY	Change to channel . . .
QSZ	Send each word or sentence more than once
QTE	What is my bearing from you?
QTH	What is your position?
QTI	What is your course?
QTJ	What is your speed?
QTL	What is your heading?
QTN	Departure time?
QTR	Correct time?
QTU	Which hours is your station open?
QTV	I will listen for you on channel . . .
QTX	Listen for me until . . .
QUA	Have you news of . . .?
QUD	Have you received the emergency signal sent by . . .?

QUF	Have you received the distress signal sent by . . .?
QUM	Is the emergency traffic ended?
QUO	Shall I search for . . .?
QUR	Have survivors been picked up?
QUS	Have you sighted survivors or wreckage?
QUT	Is the position marked?

The internationally recognized phonetic alphabet is in common use by radio enthusiasts – professional, amateur or CB – all the time, and is used frequently for spelling complicated words or in describing places, people or things by their initials:

ALPHA
BRAVO
CHARLIE
DELTA
ECHO
FOXTROT
GOLF
HOTEL
INDIA
JULIET
KILO
LIMA
MIKE
NOVEMBER
OSCAR
PAPA
QUEBEC
ROMEO
SIERRA
TANGO
UNIFORM
VICTOR
WHISKY
X-RAY
YANKEE
ZULU

Rigspeak

The CB slang is probably the thing which has attracted most attention in this country over the past few years. For some reason the majority of media outlets seem to regard the jargon as a form of mystic new language available only to acolytes of the radio waves. In consequence of this they have repeatedly devoted much valuable time and attention to the subject of CB, all of which has been of a startlingly superficial nature; it has always, it seems, been nothing more than a sly way of continually returning to the prospect of Cabover Pete and his reefer.

Quite why this should be so is far from certain or obvious; in the main the jargon is an inherited redneck/backwoods/country and western mixture of doubletalk. Far from being verbal short-hand, which is traditionally what radio transmission requires, it is more often than not longwinded, obscure and relatively pointless. Above all it is identifiably American, and though there may not be anything intrinsically wrong with being colonial in that way, there are drawbacks in this particular instance. If you'd ever heard someone say 'For sure, for sure good buddy, that's a big ten four, c'mon' in a Grimsby accent you'll know it's no joke.

Nevertheless the jargon exists. And to be honest it's early days yet, as far as British CB is concerned, and there's plenty of time for us to develop our own jargon based on our own needs and culture. To an extent this is already happening, and a rather dryer and generally more cryptic way of describing things is just beginning to emerge. It's perhaps strange that there has been little usage of our already existent and reasonably well-known Cockney slang, or the principles upon which it is established – but still, time will tell. Meanwhile you will need to know what people are on about when they're on the air (and off it – some of them talk like that all the time) and modulating like crazy. And if you mix the American slang in with a thick Glasgow accent you know you're going to need some help. For that reason there follows a brief list of fairly common words and phrases. It is by no means either complete or comprehensive, but it has the advantage of being dead common; it is also confined as far as possible to those things which you need to know. The rest of it you'll pick up as you go along. Whether or not you choose to count that as an achievement is up to you. . . .

ADVERTISING police car with flashing lights and siren
AFFIRMATIVE yes
A LITTLE HELP extra power – a linear amplifier
ANCHOR MAN base station operator also anchored modulator

BACK replying, as in COME BACK, BRING BACK etc.
BACK BENDER Sideband user
BACK DOOR last vehicle in a convoy
BACK DOOR CLOSED rear of convoy covered for police
BACKGROUND interference
BACK OFF stop transmitting or slow down
BACK OFF THE HAMMER slow down
BACKSIDE return trip
BACKSTROKE return trip
BAGGING police catching speeding motorists
BALLET DANCER antenna blowing in the wind
BAREFOOT using a rig legally, without extra power
BASEMENT channel 1
BASE TWENTY operator's home
BEAN STORE restaurant
BEAR police
BEAR BAIT speeding vehicle without CB
BEAR CAGE police station
BEAR CAVE police station
BEAR IN THE AIR police helicopter or aeroplane
BEAR REPORT information on the location of the police
BEATING THE BUSHES scout vehicle ahead of convoy going
fast enough to attract police attention and draw them from
concealment
BEAVER small furry animal with sharp teeth
BEAVER BREAKER female on channel
BEAVER HUNT searching for beaver. Once bitten, twice shy,
eh?
BETTER HALF spouse
BETWEEN THE SHEETS sleeping
BIG C North Circular Road
BIG CIRCLE North Circular Road
BIG SWITCH the on/off switch on a CB rig
BIG10–4 agree one hundred per cent
BLACKTOP major road

114

BLEEDING transmission affecting channels either side of the actual one in use

BLOOD WAGON ambulance

BLOWING SMOKE loud clear signal

BLOW THE DOORS OFF overtake

BOBTAIL artic tractor unit without a trailer

BOOTLEGGER operator without a licence

BOOTS a linear amplifier

BOULEVARD main road

BRA BUSTER lady with huge norks

BREAKER a CB user

BREAKING THE NEEDLE coming in loud and clear

BROWN BOTTLES beer

BRUSH YOUR TEETH AND COMB YOUR HAIR warning, radar ahead

BUBBLE GUM MACHINE police vehicle with flashing lights

BUBBLE TROUBLE a tyre problem

BUCKET MOUTH CB user who talks a lot on air, usually using bad language

BUMPER LANE overtaking lane on motorway

BURNER a linear amplifier

BURNING MY EARS loud clear signal

BUST getting caught by the law for speeding, illegal transmission etc.

BUTTON PUSHER CB user who presses the PTT bar but never speaks

BUZBY GPO monitors

CABOVER type of lorry which is flat-fronted

CAMERA police radar

CARRIER radio wave with no voice transmission

CHECK THE SEAT COVERS look at the passengers

CHECKING MY EYELIDS FOR PINHOLES sleeping

CHICK woman or girl

CLEANER CHANNEL channel with less interference

COME AGAIN repeat last transmission

COME ON bring it back etc.

COME BACK bring it back and so on

COMIC BOOK trucker's log book

COMING OUT THE WINDOWS good loud reception

CONVOY line of vehicles in CB contact travelling together
COPY receive a transmission
COPYING THE MAIL listening to a CB conversation without speaking
COUNTRY CADILLAC farm tractor
CUT THE COAX turn off the rig
CUT SOME Zs get some sleep
CREDIT CARD boyfriend, male passenger with female breaker

DEAD KEY person pressing the PTT bar for some time without speaking thus blanking out the channel
DEAD PEDAL slow moving vehicle
DIARRHOEA as in verbal diarrhoea – breaker who won't shut up
DIESEL DIGIT channel 19, the truckers' channel
DO IT TO ME answer, come back, come on, etc.
DON'T FEED THE BEARS don't get caught speeding
DOUGHNUT tyre, also more recently, a roundabout
DOWN AND GONE turning off the CB
DOWN AND ON THE SIDE transmission over but listening
DO YOU COPY? do you understand?
DRESS FOR SALE prostitute

EARS CB radio or antenna for same
EARWIG listen to a conversation
EASY CHAIR vehicles in the middle of a convoy
EIGHTS goodbye, but with love and kisses. Use this farewell with caution
EVEL KNIEVEL motorcycle policeman
EVERYBODY IS WALKING THE DOG channels are all busy
EYEBALL meet someone or look at something
EYETIES the Italians, who all have their own CB rigs and use them

FEED THE BEARS get a speeding ticket
FIVE as in, give me a five, transmit numbers 1 to 5 for a radio check
FIVE BY FIVE perfect signal. Also, KICKING FIVES or READ YOU IN FIVES
FLAG WAVER motorway maintenance worker

116

FLIP-FLOP return trip
FOLDING CAMERA Visual average speed computer and recorder (VASCAR)
FOR SURE agreed
FOUR as in THAT'S A FOUR means yes
FOUR TEN emphatically yes
FOUR-WHEELER car
FOXHUNT meanies looking for CB
FRONT DOOR leading vehicle in a convoy
FUNNY BOOKS porno magazines
FUR-LINED SEATCOVER female passenger

GET HORIZONTAL go to sleep
GOING DOWN switching off the CB rig
GO JUICE petrol or diesel fuel
GOLDILOCKS a mobile businesswoman
GOOD BUDDY a fellow CB user
GOOD LADY the same but a lady
GOOD NUMBERS best wishes
GOT A COPY? can you hear?
GOT MY EYEBALLS PEELED looking
GO TO 100 head for a rest-room stop
GREEN STAMPS money
GROUND-CLOUDS fog

HAMMER accelerator
HANDLE nickname by which CB user is known on air
HANG A LEFT/RIGHT make a left/right turn
HARVEY WALLBANGER reckless or bad driver
HIGH GEAR using a linear amplifier
HIT THE HAY go to sleep
HOLDING ON TO YOUR MUDFLAPS driving close behind you
HOLE IN THE WALL tunnel, area of bad reception
HOME TWENTY home address
HOTEL OSCAR the Home Office

IDIOT BOX television set
INVITATION speeding ticket or summons

JAILBAIT very young girl
JAM SANDWICH marked police traffic car
JUICE petrol or diesel fuel

KEEP THE SHINY SIDE UP AND THE GREASY SIDE
DOWN have a safe trip
KEEP YOUR NOSE BETWEEN THE DITCHES AND
SMOKEY OUT OF YOUR BRITCHES have a safe trip
KEYBOARD controls on a CB set
KEYING THE MIKE pressing the transmit button without
speaking
KODAK police radar set
KOJAK WITH A KODAK policeman operating radar

LADY BEAR policewoman (not only Angie Dickinson)
LANDLINE telephone
LAY AN EYEBALL ON meet in person, go to see
LET THE CHANNEL ROLL let others use the channel
LETTUCE money
LINEAR AMPLIFIER illegal device to boost transmit power
higher than the permitted maximum
LOAD OF VW RADIATORS empty truck or empty container
of any sort

MAKING THE TRIP getting your signal out
MAMA BEAR lady police officer
MAYDAY internationally recognized distress signal
MEANIES anti-CB authorities
MEAT WAGON ambulance
MIKE microphone
MOBILE EYEBALL check out a car or truck while moving
MOBILE PARKING LOT car transporter
MODULATE talk
MONSTER LANE nearside lane on a motorway
MOTION LOTION petrol, diesel fuel
M-20 meeting place

NEGATIVE CONTACT no answer to a call
NEGATORY no
NINE repeat last transmission as in GIVE ME A NINE

118

ON CHANNEL on the air
ONE-EYED MONSTER television set
ON THE PEG at the legal speed limit
ON THE SIDE monitoring the channel, listening watch
OPEN on the air
OTHER HALF wife, husband, girlfriend, etc.
OUT transmission completed, no further contact expected, as in FINAL
OVER transmission ended, your turn to speak
OVERMODULATING muffled or distorted signal
OVER YOUR SHOULDER behind you

PANIC IN THE STREETS area being monitored by the Post Office
PAVEMENT PRINCESS prostitute
PERMANENTLY 10–7 dead
PLAIN WRAPPER unmarked police car
PLAY DEAD stand by or refuse to answer
PORTABLE FARMYARD cattle truck
POSITIVE yes, affirmative
PREGNANT ROLLER SKATE VW Beetle
PRESS THE SHEETS go to sleep
PRESSURE COOKER sports car
PULL THE BIG SWITCH go off the air
PUT AN EYEBALL ON look at, meet
PUT THE HAMMER DOWN accelerate
PUT THE PEDAL TO THE METAL accelerate, travel flat out

Q CODE internationally recognized set of code signals
QSL CARD postcard sent to confirm CB contact, usually over a long distance
QUICK TRIP AROUND THE HORN scan all the channels for activity

RADIO CHECK report on quality and strength of transmission
RATCHET JAW operator who talks too much
READING THE MAIL eavesdropping
REEFER refrigerated trailer
RELOCATION CONSULTANT removal van
RIG truck or CB set

RINGING YOUR BELL someone calling you
ROAD TAR coffee
ROCK crystal used to tune CB rig
ROGER yes, affirmative, also ROGER DODGE, ROGER D
ROLLER SKATE production car
RUBBERBANDER novice CB operator
RUNNING BAREFOOT using legal power for CB

SAILBOAT FUEL wind or an empty petrol tank
SALT SHAKER salt spreading vehicle
SEAT COVER female passenger
SEEING EYE DOG radar detector
SEVENTY-THREES best wishes, regards
SHAKE THE TREES AND RAKE THE LEAVES vehicle at rear and front of a convoy watching for police in wait ahead or catching up from behind
SHAKING THE WINDOWS loud and clear
SHOTGUN male passenger
SIDEWINDER SSB user
SITTING UNDER THE LEAVES concealed police car
SIX-LANE CAR PARK motorway
SKATING RINK slippery road
SKY HOOK antenna
SLAMMER jail
SLIDER illegal device used to transmit between authorized channels
SMOKEY the police
SMOKEY ON RUBBER police vehicle
SMOKEY REPORT information on location of police
SMOKEY WITH A CAMERA police using radar
SMOKEY WITH EARS police with CB
SNAFU cock-up
SPAGHETTI Italian CB transmissions
SPARKS electrician
STACK THEM EIGHTS best regards
STEPPED ON YOU interrupted your transmission
STEREO coming in loud and clear
STINGER antenna, especially top-loaded variety
STROLLER operator on foot with a hand-held rig

SUICIDE JOCKEY truck driver carrying explosives or petro-chemicals
SUPERSLAB motorway
SWEET THING lady on channel
SWINDLE SHEET truck driver's log book

TAKE IT UP/DOWN move to a specified channel
TAKING PICTURES police using radar
TEN FOUR yes
TEN FOUR HUNDRED drop dead
TEN POUNDER excellent radio
THERMOS BOTTLE tanker truck
THIN weak signal
THIRTY THREE emergency signal
THIRTY TWFLVE definitely agreed – ten four three times over
THROWING transmitting
THROWING A CARRIER transmitting a carrier wave as in KEYING THE MIKE
THROWING NINE POUNDS coming in loud and clear
TOILET MOUTH breaker using obscene language
TWENTY location
TWO WHEELER motorbike

WALKED ALL OVER blotted out by another transmission
WALKING TALL coming in loud and clear
WALKING THE DOG talking skip
WALL TO WALL loud and clear, as in WALL TO WALL AND TREE-TOP TALL
WALL TO WALL BEARS police everywhere
WALL TO WALL SPAGHETTI overwhelming skip from Italy
WALLY an idiot
WHAT'S YOU'RE TWENTY? where are you?
WHEELS a vehicle, also set of wheels
WINDOW WASHER rain, as in GOD'S WASHING THE WINDOWS
WOBBLER vehicle without CB
WRAPPER the colour of a vehicle – in a blue wrapper, red wrapper etc

XYL wife, from Ex Young Lady

121

YL girlfriend, from Young Lady
YOU GOT IT the channel's yours, go ahead, agreed
YO-YO vehicle varying its speed

ZOO police station

Technical Jargon

You will doubtless find that there are a great number of words or expressions in use on CB than you are familiar with. Some of them are important, others aren't. To make life a little easier here are a few of them. Again they're only the most common ones and they're explained as simply as possible.

AC Alternating Current, as used in mains electricity
AGC Automatic Gain Control. Built in to a CB rig it maintains constant volume on both strong and weak stations
AM Amplitude Modulation – a way of putting a voice signal onto a radio wave
AMPERE Unit of electrical current
ANL Automatic Noise Limiter. Accessory on a CB rig which reduces interference from vehicle ignition and other sources
ANTENNA The aerial to which a CB rig is connected
APA Aerial Pre-Amplifier boosts incoming signal
ATU Aerial Tuning Unit – a fine tuning device

BAND A group of frequencies or channels
BASE STATION Transceiver at a fixed location
BEAM Directional type of CB antenna

CARRIER The radio wave which carries voice signals
CHANNEL Specific frequency within a radio band
CLARIFIER Control on Sideband sets which helps make the incoming signal intelligible
CO-AX Shielded cable that is used in antenna feedlines
CRYSTAL Device used to tune the CB exactly to a chosen frequency

DC Direct Current as used in motor vehicles
DECIBEL Unit of sound measurement used to rate antennas for performance
DELTA TUNE Control on CB rig for fine tuning stations slightly off channel
DUAL CONVERSION Receiver circuit to cut out interference from stations using adjacent channels
DUMMY LOAD Device which turns radio waves into heat and

allows running a transmitter without an antenna for off-air testing
DX Long-distance exchange of radio signals

EARTH Connection to the ground itself, or a common wiring
harness through a chassis in a car or radio set
ELEVEN-METRE BAND Another way of describing the
27MHz CB band. 11 metres is the wavelength of 27MHz CB

FET Field Effect Transistor – high quality transistor used in the
more expensive rigs
FIELD STRENGTH METER Instrument which measures the
strength of a radio signal, useful for tuning and antenna positioning
FREQUENCY SYNTHESIS Circuit which reduces the number
of crystals necessary for all-channel reception
FM Frequency Modulation – a way of putting voice onto a radio
carrier which is more refined than AM

GAIN An increase in power
GMRS General Mobile Radio Service
GROUND PLANE Nondirectional or omnidirectional antenna
for base station use. It has one vertical driven element and several
radial rods around the base

HANDSET Combined microphone and earphone like a telephone

IC Integrated Circuit. A one-piece chip holding thousands of
transistors. Several entire sections of a transceiver
IMPEDANCE The opposition or resistance to a flow of electric
current. 27MHz CB sets have an impedance of 52 Ohms throughout for a perfect match

JACKPLUG A single pin plug for audio connections like the
loudspeaker or PA horn

LINEAR AMPLIFIER Simple but powerful amplifier to boost
output of a CB rig to an illegal level
LOADED ANTENNA Antenna shorter than wavelength but
with a loading coil somewhere in its construction
LOADING COIL Wire wound coil used to increase the electrical length of an antenna

124

MHz MegaHertz. Measure of how many times a radio wave vibrates in a second. Mega means million and Hertz is cycles per second. A 27MHz signal therefore vibrates 27 million times per second

MICROVOLT One millionth of a volt

MIKE An abbreviation for microphone, also sometimes seen as 'mic'

MODULATING Vibrating a radio carrier with the voice

NOISE BLANKER Device similar to an Automatic Noise Limiter but more sophisticated

OHM Measurement of electrical resistance

OMNIDIRECTIONAL The sending of a signal in all directions. Also described as Nondirectional

OUTPUT POWER The power generated by a transceiver or other radio/audio device, measured in watts

PEP Peak Envelope Power is used to measure the power output of Sideband sets

PLL Phase Lock Loop. Like a synthesiser this circuit in the rig allows the transmission of signals on many frequencies without the need to have a crystal for each channel

PTT Push-to-talk bar on a radio mike allows transmission when pressed and reception when released

QUARTER-WAVE ANTENNA The basic CB antenna, which can be used for base or mobile installations

RF Radio Frequency. Electrical field which vibrates fast enough to create radio signals

RX The receive stage of a transceiver

RIG Slang word for a CB transceiver

SELECTIVITY The ability of a receiver to ignore interference and transmissions not exactly on the frequency it is tuned to

SENSITIVITY The ability of a CB set to receive weak, distant transmissions and amplify them to the point where they become audible

SIGNAL STRENGTH Strength of an incoming signal measured

on a scale of 1–9 on an S-meter

SKIP Phenomenon whereby short-wave radio signals are reflected by the ionosphere

S-METER Scaled dial on a CB rig to show strength of received transmissions

SQUELCH Control on a CB set used to eliminate noise from interference when no signal is being received

SSB Single Sideband. An efficient means of sending a radio signal which makes better use of the available channel allocations

SWR Standing Wave Radio. A way of comparing and measuring forward and reflected power in an antenna system

TX The transmit stage of a radio transceiver

TRAFFIC Radio message

TRANSCEIVER A radio set capable of transmitting and receiving radio signals

TRANSISTOR Tiny solid device which performs basic electrical switching functions

TVI Television Interference

VOLT Measurement of electric current or pressure

VOX Voice operated transmitter, which is activated by speech in the microphone rather than a PTT bar

WATT Measurement of electric power

WHIP An antenna

The first properly constituted organization which was concerned solely with CB radio was the Citizen's Band Association (CBA), which began its life in 1976. For a long time it was the only such body, but towards the close of the seventies the number of people interested in CB began to grow quite appreciably, and as they found themselves increasingly hounded by the forces of law and order they gradually began to draw together in groups, as all oppressed minorities will do.

As 1979 became 1980 their numbers began to swell at an astounding rate. Within months there were more and more clubs, and all of them were attracting larger and larger membership, as the CB pirates up and down the country joined into cohesive bodies. Very swiftly their membership became so numerous that the emphasis behind their *raison d'être* began to shift; slowly at first, and with great subtlety, but gathering force and momentum, until the gentle pressure of the tide of public opinion became a rampaging hurricane of outrage. No longer herding together simply for self-protection, the clubs and their members became the fighting cohorts of the pro-CB lobbyists, rolling irrevocably forwards behind the small group of Praetorians from the CBA and the National Committee.

While the National Committee handled all the political skirmishing with considerable skill, finesse and dedication there is no doubt at all that all their verbal swordsmanship would have achieved nothing had it not been for the vast ranks of the club members drawn up behind them. There can hardly be a single town in the country which did not have at least one CB protest march of its very own during 1980 or '81. Meetings, demos, marches, rallies, convoys; it seemed that there was one taking place somewhere in the country on every weekend in 1980.

And all the time the protest marches were happening on this vast scale the clubs were also finding time to go about their everyday business; sorting out problems with interference to various people using radio in their area, organizing their weekly meetings and newsletters, collecting subscription money for their 'bust funds', writing hundreds of letters to their own MP or Government Ministers, all kinds of things. And yet they were always there when they were needed.

When a two-year-old girl went missing in a Sussex forest one Friday night a police appeal for assistance from the public produced 50 members from the South Coast Breakers within minutes. While police officers turned a blind eye the breakers joined in the search using a number of illegal hand-held and mobile rigs to co-ordinate part of the operation, getting better results from the search line and freeing police officers to expand the area being searched. The baby was found alive on the Sunday as a direct result of assistance from the Sussex breakers, who were still actively involved in the search.

When Barbara Waite abandoned her two-year-old baby on the steps of Folkestone police station and then vanished into the night the police put out a TV newsflash at 10pm. The local breakers immediately put out their own 10–33 and by 10.15 there were more than 60 radio-equipped volunteers assisting police in their search for the missing woman. Within the hour there were more than 100 of them under the control of a single base unit, and they stayed with the search for some four hours. Mrs Waite was found alive but unconscious on private land the next morning.

Local breakers in Huddersfield responded to a 10.33 when a three-year-old child went missing and 300 of them spent hours searching, again directed by base stations, until the search ended in tragedy when the girl was found dead.

And at the same time as all this was taking place just about every CB club in the country held dances, pram races, sponsored swims, all kinds of ludicrous events, subjecting themselves to extremes of effort, exertion and, in some cases, ridicule. In every single instance the object was to raise money. None of it was spent on beer or the CB legalization campaign. Every penny went to charity. Dialysis machines, Cancer Research, Help a London Child, you name it. Groups of disabled or underprivileged kids were sent to the seaside to funfairs, on day trips in the country, the lot.

It's a great pity that there was never a central record made of all of these charitable acts and donations. The total gift in time, money and, above all, care, must be truly staggering. Someone

once said that CB was a nice way to meet nice people. Now tell me he was wrong.

At the end of this chapter you will find a whole long list of club addresses. There's almost bound to be one close to you. If there isn't you must be living in the wrong place. Just give a shout on your calling channel and you're bound to turn up a local club eventually, more than likely two. If you don't you'd better start your own. You'll be in very, very good company.

There are other ways to make friends in the CB world if you don't like social gatherings or have trouble getting about (although I'm sure your local club could get you to their meetings if they knew you were there and wanted to attend). Probably the easiest and most rewarding is this business of QSL cards. The International Q-code provides under the code QSL the explanation 'please acknowledge receipt'. Among Radio Hams the practice of sending a postcard to confirm radio contact with each other, particularly over long distance, is well established. It is perfectly natural for habitual users of the Q-code to call such confirmations QSL cards.

Although they were primarily very basic communications many Hams designed their own cards and had them specially printed simply in order to add a rather more personal touch to the whole affair. In recent times this habit has been seized upon with great enthusiasm by CB users as well. Generally speaking the range of a 4-Watt CB rig does not allow very much in the way of long-distance communication except in very unusual circumstances. Thus the practice of sending out QSL cards is now regarded as an end in itself rather than as merely a facet of a radio link.

The idea is simple. Design a card yourself (or let someone do it for you) which bears some relationship to your CB handle and then zap it off to someone else interested in QSL. Don't send it on its own though; send several. The recipient will then distribute them among his own QSL contacts and pretty soon you'll have QSL cards arriving by every post, including airmail. The QSL hobby is now self-sufficient, somewhat independent of the radio facility which gave birth to it, and very international indeed. There are now several QSL clubs, both domestic and overseas, which anyone may join by sending a small sum and a number of their own QSL cards, which are remarkably good value. The 'multiplier' effect of joining such a club produces a breathtaking number of QSL cards and opens the door to a wealth of new friends.

It is, you must realize, a time-consuming operation. If you are to participate fully in the hobby then you must QSL 100 per cent, which means returning the compliment every time you receive a new card. But in the few short months that the hobby has been gaining attention in this country many British QSL fans have developed friends in all parts of the world, sometimes in the most unlikely places. Anyone with interests of a vaguely postal or philatelic disposition, or with plenty of spare time, would have to look a long way to find a hobby as rewarding and entertaining as QSL.

In conjunction with Britain's best-known exponent of the QSL art, Mike 'The Medicman' Newbold, we have compiled a list of QSL clubs which have proved themselves honest, reliable and interesting. Make the most of them.

British CB Clubs

27 Club
Every Monday at
Saxon Tavern
Southend Lane
Catford SE6

A78 CBBC
c/o 117 Main Street
Largs
Ayrshire
Scotland

Aire Valley Breakers Club
53 Albert Road
Saltaire
Shipley
West Yorks

Airwave Breakers
Every Monday at
The Plough
Bedminster
Bristol

Anglia Breakers Club
c/o Great White Horse Hotel
Tavern Street
Ipswich
Suffolk

Appletart Breakers Association
c/o 45 Heyers Avenue
Horley
Surrey

Aquae Sulls
c/o The CB Centre
Chelsea Road
Weston
Bath
Tel: 0225 333379

Association of Maghull Breakers
c/o 69 Vetch Hey
Netherley
Liverpool
Merseyside

Attic Breakers Club
12 Jameson Road
Bridlington
East Yorks

Australian International QSL
Swap Club
PO Box 855
Freemantle 6160
Western Australia

Avanti Breakers Club
c/o Post Office
Newton of Falkland
Cupar
Fife

Back Road Breakers
Alternate Thursdays at
Liberal Club
Garstang
Nr Preston
Lancs

Barley Breakers Club
Alternate Wednesdays
Barley Shief
New George Street
Plymouth
Devon

Barnet Breakers Club
Meet at British Legion Hall
Brookhill Road
East Barnet
Herts

Barrier Breakers
Every Wednesday at
The Railway Hotel
Netherfield
Nottingham

Barwell CB Radio Club
5 Mayfield Way
Barwell
Leics

Beachcombers Breakers
Association
c/o 3 Thursby Road
Highcliffe
Christchurch
Dorset BN23 5PA

Beech Breaker's Club
Every Sunday evening at
Blacksmiths Arms
Thornwood Common
Nr Epping

Big C Club 80
c/o Dominix
PS 14
The Market
Carmarthen
Dyfed
S Wales

Big Eyeball Breakers
Every Thursday at
The White Hart
Devonshire Hill Lane
London

Big Top Breakers Club
5 Council Villas
Melton Ross
Barneteby
South Humberside

Big Wheelers Association
35 Alexander Court
Lansbury Park Est.
Caerphilly
Mid Glam.
Wales

Biscuit Town Breakers
PO Box 123
Reading
Berks
(send SAE)

Blackpool Breakers Club
c/o ADS Electronics
239 Dickson Road
Northshore
Blackpool

Boomerang Breakers Club
Meet at White Lion
Moulton
Northampton

Border Breakers Club
c/o Cathedral Garage
Weybread
Harleston
Norfolk

Boston Breakers Club
19 Pool Lane
Kinson
Bournemouth
BH11 9DX

Bottle City Breakers
31 Farm Road
Clock Face
St Helens
Merseyside

Bottsford CB Association
c/o 8 Spusby Road
Scunthorpe
South Humberside

Boulevard Breakers Club
56 Kirkdale Drive
Glasgow GS2 1ET

Bourne End Breakers Association
Fridays at
The Fire Fly Pub
Bourne End

Bournemouth Independent
Breakers Association
Every Tuesday
Coach House Motel
Ferndown
Dorset

BP Breakers Association
19 St Helens Avenue
Flimby Maryport
Cumbria

Bracknell Breakers
Every Sunday at
The Bridge House
Wokingham Road
Bracknell
Berks

Braithwell Rig & Twig Club
Every Tuesday
Braithwell WMC
South Yorkshire

Bramley and District Breakers
5 Ferncliffe Terrace
Leeds
Yorkshire

Breaker One Four Club
c/o OK Corral
Napier Barracks
BFPO 20
West Germany

Breakers Town CBC
Every Thursday
c/o The Stanley Club
Stanley Road
Carshalton
Surrey

Breakers Yard CBC
Every Monday at
St Helier Arms
Carshalton
Surrey
Tel: 01-669 5441

Bricket Breakers Club
c/o Watford Component Centre
7 Langley Road
Watford
Herts

Bridgetown Breakers Club
Meet at Phoenix Social Club
Heesle Road
Hull

Bristol Breakers
120 Beaufort Road
St George
Bristol 5

Bristol CBC
1A St Peter's Rise
Headley Park
Bristol
BS13 7LU

Brook Breakers CB Club
Every Wednesday at
Badger in the Brook
Shirebrook
Mansfield

Broadland Breakers Club
First Sunday at
White Swan
Stalham
Norfolk

Brown Bottle Breakers
c/o The White Horse
Norton Road
Thelnetham
Diss
Norfolk

Buckinghams Breakers Club
35 Addington Road
Buckingham

Bruggen Bandits CB Club
On channel 13
BFPO 25

Burns Breaker Club
c/o Braehead Hotel
Whiltletts Road
Ayr
Scotland

Bury CBC
c/o CB Paradise
69 Northgate Street
Bury St Edmunds
Suffolk

Byron CB Radio Club
Byron Hotel
Ruislip Road
Greenford
Middlesex

Caketown Breakers Club
Every Sunday at
The Queens Hotel
Pontefract

Campaign for 27MHz AM CB
Radio
10 Lochnell Road
Dunbeg
Connel
Argyll PA37 1QJ

Canary City Breakers Club
Meet first Tuesday of month at
Ebenezers Freehouse
Salhoouse Road
Norwich

Canyon Breakers Club
Meet every Wednesday at
Hillstown Miners Welfare
Hillstown
Chesterfield

Cardiff & District Breakers
12 Aberdored Road
Gabalfa
Cardiff

Carlton & Langold United
Breakers
Thursdays at
Langold Hotel
Langold
Worksop
Notts

Carrick CB Club
Tuesday night at
Carrick Hotel
Maybok
Ayrshire

Castle Breakers
c/o Rose & Crown
High Street
Tonbridge
Kent

CB 007 Breakers Club
Tuesday nights at
The Windmill Club
Rotherham

CBA Central Scotland
5 Carronvale Avenue
Larbert
Stirlingshire

CBA
Coronation Service Station
Middleton Road
Heywood
Lancs

CBA Fife
16 Bayview Crescent
Methil

CBA Reading
PO Box 123
Reading

CBA Sussex
15 Buckingham Mews
Shoreham By Sea
Sussex

CBCB Club
103 Southwood Road
Downside
Dunstable
Beds

CBGB
CB House
Crosby
Liverpool

CB Information Centre
7 Sandringham Crescent
Harrow
Middx HA2 9BW

CB–NE
PO Box 61
Sunderland SR3 1EZ

CB Radio Action Group
55 Dartmouth Road
Forest Hill
London SE23

Central 27 Breakers Club
Alternate Tuesdays at
The Bruce Inn
Nr Landmark
Springkerse Road
Stirling

Central England Breaker's
Association
Meet on Tuesdays at
Staffordshire Volunteer
Collingwood Road
Bushbury
Wolverhampton

Cheesy Breakers Club
116 St Christopher's Drive
Caerphilly
Glamorgan

Cheltenham Breakers Assn
6 Pitville Crescent
Cheltenham
Or: The Crown and Cushion
Bath Road
Cheltenham

Chichester & District
Breakers Club
Every Sunday at
Bulls Head
Fishbourne
Chichester
Sussex

China Town Breakers Club
c/o 54 Oxford Road
Penkhull
Stoke on Trent

Circle C Breakers
c/o The George Hotel
Crewkerne
Somerset

City Circle CB
Bedford Green
Horseferry
Leeds

Clean Air Association
Mondays at
Woodhays pub
Wednesfield
Wolverhampton

Clog Town Breakers Club
33 Pendle Court
Astley Bridge
Bolton BL1 6PY
Tel: Bolton (0204) 50046

Clogtown Claypit Breakers Club
Aquarious Club
Halden Street
Dean
Bolton

Club 14
Every Thursday at
Spotted Cow
Willesden High Road
NW10

Club Breakaway
c/o 123 Hasler Road
Canford Heath
Poole
Dorset

Clyde Coast Breakers
c/o Island Hotel
New St
Stevenston
Ayrshire

Clydeside Breakers
Supporters Club
62 Rosemount Crescent
Carstairs
Lanarkshire

27 Coastline Breakers
PO Box 24
Rhyl
Clwyd
North Wales

Coastline Breakers Club
Meet every Wednesday at
Wash and Tope
Le Strange Terrace
Hunstanton
Norfolk

Copy Cats Club
The Manager
Martholme Grange
Altham
Accrington
Lancashire

Cottonmount Breakers Club
Alternate Tuesdays at
Cottonmount Arms
Mallusk
Newtownabbey
N. Ireland

County Area Breakers Club
4 Corbert Gardens
Ardersier
Inverness

Country Town Breakers Club
Angorfa
Baptist Street
Penygroes
Caernarfon
Gwynedd

Crewe Breakers Club
1 Main Road
Crewe
Cheshire

Cromwell Breakers
c/o The Winning Post
Market Deeping
Peterborough

Cuckooland Breakers Club
PO Box 2
Penicuik
Scotland

Dare Breakers Club
c/o Paul Venn
72 Tre Telynog
Cwmbach
Aberdare
Mid Glam
S Wales

Delta Breakers
Sundays at
Invicta Co-op Sports Club
Burnham Road
Dartford
Kent

Derwent Valley Breakers
10 Prospect Terrace
New Kyo
Stanley
Co Durham

Deveron Valley Breakers Club
Turriff
Aberdeenshire AB5 7PQ

Diamond Breakers Club
Diamond Jubilee Club
South Kirby
West Yorkshire

The Diamond Breakers Club
c/o 16 West Close
Stevenage
Herts

Didcot and District 27 Club
Meet first and third Sunday
of the month at
The Rio Hadden Hill
Nr Didcot
Oxfordshire

Ditch Breakers
Meet every Monday at
The Railway Hotel
Netherfield
Nottingham

Dinnington & District
Breakers Club
Alternate Wednesdays at
The Squirrell
Dinnington

Dixieland Breakers
PO Box 25
Grimsby
South Humberside

Doctor's Cure Breakers
Meet in the Legion
Healing
Grimsby

Don Valley Breakers
15 Roseberry Avenue
Hatfield
Doncaster

Dorset Nob Breakers Club
c/o 3 Barr Lane
Burton Bradstock
Bridport
Dorset

Dragon Breakers Association
c/o 70 Ffordd Lligwy
Moelfre
Anglesey
Gwynedd

Driffield CB Association
c/o 22 Haworth Walk
Bridlington
East Yorkshire

Dukesville Breakers Assn
23 Potter Street
Worksop
Notts

Eagle Breakers Club
c/o 76a Penn Hill Avenue
Parkstone
Poole

EarthQuake City Breakers Club
DL 71
Rotherham Record
Regent House
Rotherham

East Antrim CBRC
PO Box 4
Antrim

East Coast Breakers Assn
c/o The Tartan House
Frating
Nr Colchester
Essex

Eastern Counties Open
Channel Club
c/o Everards Hotel
Cornhill
Bury St Edmunds
Suffolk

Edinburgh Breakers Club
Meet Monday at
Sinatra's Lounge Bar
St James Centre
Edinburgh
Scotland

Edinburgh CBRC
22 Rose Gardens
Edinburgh EH9 3BR

Elite Breakers
The Father Thames
Albert Embankment
London SE1

Essex Citizen's Band Club
24 Bryony Close
Witham
Essex CMB 2XF
Tel: Witham (0376) 513532

Falkirk and District Open Channel
PO Box 15
Falkirk
Scotland FK1 1AA

Farnborough Area Breakers
Every Thursday
The Oasis Club
Alexander Road
Farnborough
Hants

Fish Town Sea Bees
Meet at Big Wheel
Grimsby
South Humberside

Five Bridges Breakers Club
c/o Barfield
Oakville Road
Hebden Bridge
West Yorkshire

Flixton, Urmston & Davy Hulme
Good Buddies Assn
PO Box 2
164 Corn Exchange Buildings
Manchester 4

Forfar and District Breakers Club
Meet every Monday in
Stag Hotel
Forfar

Freedom Breakers International
11 June Street
Bootle
Liverpool
Merseyside

Frog and Nightgown Breakers
Club
c/o Amberwell
Pottersheath Road
Welwyn
Herts

Give Us A 9 Club
Every Tuesday at
Hand in Hand
Boxhill
Surrey

GK 13
D5130 Geilenkirchen
West Germany

Glasgow CBC
361 Hallhill Road
Glasgow G33 4RY

Golden Gate Breakers Club
c/o The Deva Restaurant
Cliff Road
Dovercroft
Harwich
Essex

Good Buddies Club
Alternate Wednesdays at
Halfway House Hotel
Kingseat
Fife

Gower Breakers Club
PO Box 12
Swansea
South Wales

Grampian Breakers Club
59 Jasmine Terrace
Aberdeen
Scotland

Granite City CB Club
92 Forest Avenue
Aberdeen
Scotland PH 322073

Grantham Breakers Assn
8 Parklands Drive
Harlaxton
Grantham
Lancs

Grass Court Breakers Club
Every Sunday at
The Wagon & Horses
Haydock
St Helens

Guildford City Breakers
Tuesdays at
The Cannon
Portsmouth Road
Guildford
Surrey

Gwent Breakers Club
Meet every Wednesday at
The Gladiator
Malpas
Newport

Hangmans Breakers Club
Meet on Thursdays at
The Staffordshire Knot
Birmingham Road
Wolverhampton

Harrow and Wembley CB Group
26 Greenway
Kenton
Middlesex

Hazzard County Breakers
c/o Oakshaw Hall
School Wynd
Paisley
Renfrewshire

Hazzard County Breakers Club
22 Radcliffe Avenue
Chaddesden
Derby

Hazzard County Breakers Club
Meet first Tuesday of month at
Fleet Country Club
Surrey

Heart of Oak Breakers Club
Feltham Road
Ashford
Middlesex

Hereford 14 Club
Meet every Monday at
Crystal Rooms
Hereford

Hereward Breakers Club
17 Munton Fields
Ropsley
Grantham
Lincs

Herts Citizen Band Radio Assn
c/o Stratford Arms
Mutton Lane
Potters Bar
Herts

Highland Breaker Club
PO Box 39
Inverness

Hillbillys
Meet every Friday at
The Friend at Hand
West Wycombe Road
High Wycombe
Bucks

Hornblower Open Channel Club
c/o 65 North Street
Ripon
North Yorkshire

Hucknall Welfare Breakers Club
Hucknall and Linby Miners
Welfare
Portland Road
Hucknell
Nottingham

Invitation Breakers Club
Every second Sunday
The Yorkshire Dragon
Maltby
Cleveland

Ironstone Breakers Club
c/o 16 Lunedale Road
Scunthorpe
South Humberside

Journeys End Breakers
Alternate Thursdays at
Escrick Social Club
York

Junior Breakers Club
Scout HQ
Clifford Bridge Road
Coventry
1st Thursday every month

Kent and Essex Breakers
Association
Every Tuesday at
Orsett Hall
Orsett
Essex

Kent and Essex Breakers
24 Mill Lane
West Thurrock
Essex

Kings Lynn Breakers Club
c/o Cellar Man
Victoria P.H.,
John Kennedy Road
Kings Lynn

Kings Norton CB Club
Poste Restante
GPO
Lisburn
Northern Ireland

Kintyre Breakers Club
Sudown
Tarbert
Argyle

LA Breakers
Unit 13
Carlton Industrial Estate
Hawthorn Avenue
Hull

Laker Town Breakers Club
Every Tuesday at
The Cornish Man Hotel
Wythenshawe
Manchester

Lazy K
Lima Kito Radio Club
PO Box 55
Portadown
Northern Ireland

Leapool Breakers Club
c/o Maid Marion Hotel
Coppice Road
Arnold
Nottingham

LEBC (Castle Breakers)
Pete Beilby
c/o 189 Derby Road
Long Eaton
Nottingham

Leicestershire CBers
c/o Modern Motoring
68 Narborough Road
Leicester LE3 0BR

Lennox Breakers Club
4 Lismore Crescent
Oban
Argyll

Leslie Breakers
Mondays at
The Leslie Arms
Cherry Orchard Road
Croydon

Lima Bravo DX Group
PO Box 11
Oban
Argyle
Scotland

Lincolnshire, Nottinghamshire,
Derbyshire and Yorkshire area
Committee (LDNY)
8 Sunnyside
Worksop
Notts

Log Breakers
Every Monday at
Log Cabin
Royal Oak Pub
Watnall
Notts

Lorn Breakers
PO Box 11
Oban
Argyle

Lost County Breakers
Meet every Tuesday at
Pete's Paradise
Windermere
Cumbria

Lowestoft & District
Jolly Breakers
10 Viburnum Green
Lowestoft
Suffolk

Maidenhead Official Breakers
Thursdays at
The Prince Albert
King Street
Maidenhead

Mansfield Area CB Club
c/o James Maude Social Club
Forest Road
Mansfield
Notts

Market Town Breakers
PO Box 2
Ashford
Kent

MCBRA
85 Allens Lane
Pelsall
Walsall
West Midlands

Medway Breakers
55 Playstool Road
Newington
Sittingbourne
Kent

Meon Valley Breakers
4 Lawrence Road
Fareham
Hants

Merseyside 27 Club
34 Micklefield Road
Liverpool 15

Mexico City Breakers
The Old Masons Arms
High Street
Mexborough
Yorks

Mid-Kent CBC
c/o Ten Four Telecom
22 The Broadway
Maidstone
Kent

Midlands CBRC
Unit 2
72 Oval Road
Erdington
Birmingham

Midlands CB Radio Club
85 Allens Lane
Pelsall
Walsall
West Midlands

Milktown Breakers
Meet every first Sunday
in the month at
Vanity Fair
Bradford Road
Huddersfield

Molesey Open Breakers
c/o Royal Oak
337 Walton Road
East Molesey
Surrey

Monklands Breakers Club
c/o 78 South Commonhead
Avenue
Airdrie
Lanarkshire
Scotland

Moray CB Breakers Club
c/o Gearchange
40-42 Moss Street
Elgin
Morayshire

NACB
Every Thursday at
The Commodore International
Nuthall Road
Nottingham

National CB Union
PO Box 123
Reading
Berkshire

National Committee for the
Legislation of 27MHz CB Radio
47b Stoneygate Road
Narborough
Leicester

National Federation of Licenced
Breakers
142 Luttlerworth Road
Nuneaton

Nationwide Breakers Club
Tentercroft Street
Lincoln

New City Breakers Club
9 St Leger Court
Linford Local Centre
Gt Linford
Milton Keynes
Bucks

New Forest CB Club
12 Westcot Road
Holbury
Hampshire

Newton Breakers Club
Meet every Tuesday at
Book and Candle
Redditch

Newtown Breakers Club
c/o 14 Cornbrook
Holland Moor 2
Skelmersdale
Lancs

Noisy City Breakers
Every Wednesday
Flamingo Night Club
Darlington
Co Durham

North Bucks Breakers
The Folly Inn
Adstock
Buckingham

North East Derbyshire 10–4 Club
c/o The Shoulder of Mutton
Hardstoft
Nr Pilsley
Chesterfield
Derbyshire

North London Breakers
Wednesday at
The Sparrowhawk
Glengall Road
Edgware
Middlesex

North Notts Breakers
4 Farm Grove
Theivesdale Lane
Worksop
Notts

Northampton Breakers Club
Wednesdays & Sundays at
The Needle
Northampton

North Sea Breakers
c/o 27 Zena Street
Glasgow

North West Breakers Association
c/o 8 Longhill Walk
Moston
Manchester M10 9NT

Norwich Social Breakers Club
72 Silver Road
Norwich
Norfolk NR3 4TD

Open Channel CBC
17 Coronation Street
Preston

Open Channel Citizens Band Club
17 Coronation Street
Blackburn

Out of City Breakers
Every Tuesday
Southall Working Mens Club
Beighton
Sheffield

Over Wyre Breakers
Fernhill Hotel
Park Lane
Preesall
Nr Blackpool

Pendle CB Supporters Club
110 Barkerhouse Road
Nelson
Lancs

Pennine One Nine Club
29 Legrams Avenue
Lidget Green
West Yorkshire BFD7 2PP

Pirates
Meet every Thursday at
The Mill
Halfway
Sheffield

Plaistow Breakers Club
Every Monday
Phoenix Club
Grange Road
London E13

Popular Breakers Club
29 Puttenham Road
Sherfield Park
Chineham
Basingstoke

Pudsey MF Citizen
Band Radio Club
54 Harley Drive
Swinnow
Leeds

Purbeck One-Nine Club
Every other Tuesday at
The New Inn
Church Knowle
Dorset

Quaker Breaker Club
c/o Waggon and Horses
East Street
Saffron Walden
Essex

Quiet Breakers Club
8 Wedgewood Road
Cheadle
Stoke on Trent
Staffs

Rainbow Breakers
c/o PO Box 56
Cookstown
Co. Tyrone
N. Ireland

R & B Club
PO Box 4
Stranraer
Scotland

REACT UK
142 Lutterworth Road
Nuneaton
Warwickshire

Red Cat 14 Breakers Club
c/o Red Lion
Derby Road
Sandiacre
Derbyshire

Redditch Area CB Club
88 Heronfield Close
Churchill
Redditch
Worcs

Redhill Radio Breakers Club
c/o The Ram Inn
Mansfield Road
Redhill
Nottingham

Rhine Cuppers CB Club
On channel 13
BFPO 40

Rhondda Breakers Club
35 Shady Road
Gelli
Rhondda
Mid Glamorgan

Rhythm and Blues Club
The Bell Hotel
Botesdale
Nr Diss
Norfolk

Richmond & District Breakers
Friday nights at Black Horse
Richmond

Ringway Sideband Club
Every Thursday at
Benchill Hotel
Wythenshawe
Manchester

River City Breakers Club
c/o 38 Worcester Road
Burnham-on-Crouch
Essex

River Exe Breakers
c/o 149 Withycombe Village Road
Exmouth
Devon

Riverside Breaker's Club
Every Friday at the
Redcroft Hotel
Bo'ness
West Lothian
Scotland

Riverside Breakers
c/o 1 St Lukes Grove
Humberstone
Grimsby
South Humberside

Road Apple DX Club UK
SAE to Robert RA 68
c/o Top Ear
London Road
Eaton Socon
Hunts

The Rolling Stones
Breaker's Club
The Moss Cottage
Nottingham Road
Ripley
Derby

Roman City Breakers Club
c/o 29 Kelston View
Whiteway
Bath
Avon

Roman Road Breakers
Tuesdays at
Galway Arms
Harworth
Nr Doncaster

The Rooftop Breakers Club
Every Wednesday at the
Gondola
Ballon Woods
Nottingham

Royal T Breakers Club
c/o 3 Manse Street
Tain
Ross-shire

St Neots Breakers Club
Every Thursday at
St Neots Working Mans Club
Hardwick Road
Eynesbury
St Neots

Saddleworth Breakers Club
Every Monday at
Well Lit Pub
Saddleworth

Sandwell Area CB Club
4 Baldwin Close
Twidale Warley
West Midlands

Saundersfoot and District
Breakers Club
14 Ryelands Place
Kilgetty
Dyfed SA68 0UX

Sedgefield Breakers Club
c/o 4 Pine Ridge Avenue
Sedgefield
Co Durham

Seven Towers CBC
15 Carnduff Drive
Ballymena
Co Antrim

Severn City Breakers Club
c/o PO Box 2
Shrewsbury

Singing Wheels CBC
c/o 2 Grenofen Cross
Tavistock
Devon PL19 9ER

Sheaf Valley BC
c/o 27 Ashberry Gardens
Sheffield

Slab Town Breakers Club
Meet every Thursday at
East and West Ardsley Social Club
Morley
Nr Leeds
West Yorks

Southend and District Breakers
Every Thursday at
Rascals Disco
Southend

South Birmingham
CB Club
Meet fortnightly at
Solihull Civic Hall
Solihull
Birmingham

South Somerset Breakers
c/o 19 Vincent Street
Yeovil
Somerset

South Wales Action Teams
16 Lanelay Park
Talbot Green
Pontyclun Mid Glam

South Wales Big 10–4 Club
139 Manselton Road
Manselton
Swansea

South Wales 10–100 Artists Club
Tuesday nights at
The Landing Strip
Swansea

South Wales Federation
of Breakers
c/o 74 Beech Court
Gilfach
Bargoed
Mid Glamorgan

South West Lancs Breakers Club
c/o 14 Cornbrook
Holland Moor 2
Skelmersdale
Lancs

Stag Town Breakers Club
Every Thursday at
Courtlands Social Club
Thorpe Road
Bellamy Road Estate
Mansfield
Notts

Steel City CBC
282 Eccleshall Road
Sheffield S11 8PE

Steeltown Breakerways
c/o 12 Keelby Road
Scunthorpe
South Humberside

Stour Valley Breakers
c/o The Red Lion
South Street
Manningtree
Essex

Stourport-On-Severn Breakers
Club
Every Sunday at
The Old Anchor
Stourport

Studley Breakers
c/o Studley Arms
Studley Green
High Wycombe
Bucks

Summer Wine Breakers
Every other Thursday at the
Burnlee Working Men's Club
Holmfirth

Swindon CB Club
23 Affeck Close
Toot Hill
Swindon

Tango Foxtrot Charlie
International DX Group
PO Box 14
Heywood
Lancs

Tayside CB Club
c/o 271 Fintry Drive
Fintry
Dundee

Telford CBC
Tel: Telford 603474

Test Valley Breakers Club
PO Box 27
Andover
Hants

Thames Area Breakers
c/o 81 Villas Road
Plumstead
SE18

Thamesdown Transceivers
Every Monday
Swindon Town Football
Supporters Club

Toadtown Breakers Club
Meet every Sunday at
Bridgend Inn
Howey
Nr Llandrindod Wells

Tunbridge Wells CB Association
Monday evenings at the
Robin Hood
Tunbridge Wells

TWINS
c/o 5 Nuthatch Drive
Earley
Reading
Berks

UK International Radio Group
PO Box 13
Long Eaton
Nottingham

Untouchables
299 Manchester Road
Kearsley
Bolton
Lancs

Walsall CB Radio Club
c/o 6 Central Close
Bloxwich
Walsall
West Midlands

Waterbabies Breakers Club
c/o 17 Furzey Road
Upton
Poole

WD40 Club
PO Box 13
Weymouth
Dorset

Wellingborough Breakers Club
Wednesdays at
Dog and Duck Pub
Wellingborough

Wessex Open Channel Club
48 Holsom Close
Stockwood
Bristol BS14 8LX

West Glamorgan Breakers
Association
25 Plas Newydd
Bagian Moors
Port Talbot
West Glamorgan
South Wales
Meet on Wednesday nights at
125 Club, Port Talbot

West London Breakers
Tuesdays at
White Hart
Southall

West Glamorgan Breakers
c/o Dock Hotel
Aberavon
Port Talbot
West Glamorgan

West London Breakers
Meet at the Steam Packet by
Kew Bridge

Weston Breakers Club
33 Lower Church Road
Weston Super Mare
Somerset

Wetherby District Breakers
9 Norfolk House
Wetherby
West Yorks

Weymouth CBC
Flat 1
39 St Thomas St
Weymouth
Dorset

White Cliff Breakers Association
PO Box 13
Dover
Kent

Wickrath Breakers
On channel 14
4050 Monchengladbach 4

Wirral CB Assn
Meet every Monday at
Riverside Restaurant
New Brighton

Woking Centre Breakers
c/o Jovial Sailor
Ripley
Surrey

Worth Valley Breakers
c/o 8 Carlisle Street
Parkwood
Keighley
Yorks

Wye Valley Breaker's Club
c/o CB Centre
106 East Street
Hereford

Wyre Forest Breakers
19 Chawson Pleck
Chawson Estate
Droitwich

152

Yorkshires Elite Breakers
Fairway Inn
Birley
Sheffield

Young Breakers Assn
22 Romley Crescent

Bolton
Lancs

Zebra County Breakers
Post Office
Sible Hedingham
Essex

QSL Clubs

AMERICA

Alaska Blue Canoe QSL Club
PO Box 3017

Australian International QSL
Swap Club
PO Box 855
Fremantle
Western Australia 6180

BELGIUM

Arenberg Swap Club
PO Box 33
3030 Heverlee
Belgium

CANADA

Telephone City QSL Club
Box 1971
Brantford
Ontario N3T 5W5
Canada

Thistle QSL Club of Canada
PO Box 4
Postal Station 'C'
Winnipeg
Manitoba
R3M 3S3 Canada

Top Dog QSL Club of Winnipeg
1139 Notre Dame Avenue
Winnipeg
Manitoba
R3E ON4 Canada

DENMARK

Scandinavian Skippers QSL Club
Krogen 3
8900 Randers
Denmark

FRANCE

Le Gaulois QSL Club
PO Box 714
26007
Valence
France

FINLAND

Tampere Radio Club
Ryydynkatu 64
SF 33400 Tampere 40
Finland

GERMANY

Berliner Bear
Postbox 2923
6750
Kaiserslautern 1
W. Germany

Super Stinky QSL Club
Postfach 2664
D6750 Kaiserslauten
W. Germany

World Amateur Group
PO Box 1243
5439 Rothernback Ww
W. Germany

GREAT BRITAIN

Big Ben DX QSL Club
PO Box 14
Godalming
Surrey
GU7 1PS

British Bulldog International
23 Russell Avenue
Colwyn Bay
North Wales LL29 7TR

British Concorde International CB
QSL Club
187 Walton Road
East Molesey
Surrey KT8 0DY

English International DX Club
225 Arnold Street
Boldon Colliery
Tyne & Wear
SR3 1EU

Scottish DX QSL Club
45 Seedhill Road
Paisley
Renfrewshire
Scotland

Voice of Scotland International
DX Club
PO Box 29
Kilmarnock
Scotland

GREECE

VIP Club of Greece
PO Box 19
Athens
Greece

HOLLAND

Royal Dutch CB QSL Club
PO Box 2744
5902 MA Venlo
Netherlands

Silly Tower QSL Club
Narcissenstr 52a
3073 cp Rotterdam
Netherlands

ICELAND

Icelandic International DX QSL
Club
PO Box 10040
Reykjavik 130
Iceland

INDONESIA

Barong Bali International DX
QSL Club
53 Denpasar
Bali
Indonesia

ITALY

Amateur Radio Italian Club
PO Box 13
Ciampino Airport
Rome 00040
Italy

Red Devil International QSL Club
PO Box 20
20079 S. Angelo
Lodigiano
Italy

NORWAY

Norway Amateurs DX QSL Club
PO Box 64
N 4030 Hinna/Stavanger
Norway

NEW ZEALAND

Gumboot QSL Club
PO Box 4127
New Plymouth NZ

Kia Ora QSL Club
PO Box 630
Wellington NZ

New Zealand and Worldwide
QSL Swap Club
PO Box 83 020
Te Atatu South
Auckland NZ

SWEDEN

Saturnus QSL Swap Club
PO Box 173
S-441 Alingsas
Sweden

Three Vikings QSL Club
PO Box 3021
Angered 3
S-424 03
Sweden

SWITZERLAND

Walkie Talkie QSL Club
Box 117
Zurich CH 8037
Switzerland

WEST INDIES

West Indies QSL Club
46 Branch Avenue
St Johns
Antigua
West Indies

During the years of campaigning for legal CB in this country there were always two arguments which carried the most weight. The first was the question of civil liberty and the right to as much freedom of speech by radio as by any other method, and the other was the direct benefit to society realizable from a properly working CB facility.

This latter subject raised issues as emotional as all the others put together. The Americans had been living with CB for a long time, and had plenty of opportunity to study and evaluate its uses. Their conclusions, reached before the wrangling on the subject even got off the ground in the UK, indicated that CB was of immense value to the community.

This conclusion was reached following a lengthy investigation into the way people were using their CB and the possible ways in which it could be used. Ultimately the benefits of a social nature – allowing people housebound for any reason a new circle of friends and similar – are beyond measurement. There is no way to calibrate loneliness or the effects of its relief. The only basis on which CB could be evaluated was at an entirely practical level where it was able to provide direct and calculable assistance to people in need. It is no surprise to find that the section of society which derived the most benefit from CB was the motorist.

The FCC found that CB was actually capable of saving life on the roads which otherwise would have been lost, provided it was correctly handled. As a result it became Federal policy to encourage and promote the use of CB radio as an aid to road safety. At the time, the possibility of roadside radio stations, with very limited range, to broadcast weather reports, traffic information and warnings of incipient danger to passing vehicles was also under review. It was concluded that very few motorists would actually spend money on a device which was capable of nothing except maybe making their motoring life safer. On the other hand they were more than willing to part with similar sums of cash for CB radio sets, since these provided them with much more than just prophecies of imminent doom.

Part of the reason for the success of the CB facility in these tests was the existence of a nationally agreed emergency channel – nine. This channel was (and still is) kept free at all times for everything except the transmission of messages concerned with emergencies; these are deemed to be situations in which life or property is threatened with damage or destruction.

Having a national emergency channel is no use at all unless someone is out there listening, obviously enough, and the results of the test were much aided by the fact that most law enforcement agencies and emergency services in the USA tend to maintain some form of listening watch on channel nine. They do, however, occasionally have other things to attend to, and it is at this point that the whole thing would have collapsed, since even if the channel is left unattended for a minute its effectiveness is destroyed. However there exists in the USA an organization rejoicing in the name of REACT, which is an easier way of saying Radio Emergency Associated Citizen's Teams. This all-volunteer group maintains a 24-hour listening watch on channel nine on a nationwide basis and its highly-skilled operators are always ready to deal with every emergency from a potentially dangerous freeway breakdown to a full-scale major disaster like an air crash.

REACT monitors can summon via landline or, more recently, via a complex variety of sophisticated radio equipment any kind of emergency assistance in less time than it would take the man on the spot to locate an operational telephone. Since its formation in 1962 REACT has handled more than 60 million distress calls, over 15 million of which have concerned traffic accidents. It was in no small way due to the existence of REACT that the Federal Government was so pleased with the results of their CB survey.

If the CBA was the first CB-related group to be formed in this country then there is little doubt that REACT was the second. To begin with it was known as The REACT (UK) Supporters Club and in this guise its hands were more or less tied. In order for it to function as successfully as it does in America REACT required a close working relationship with the authorities, particularly the emergency services. The opinion of its officers was that becoming involved with the illicit CB pirates working 27MHz AM in the UK at the time would seriously prejudice the future of any such relationship. It was therefore necessary for them to turn a deaf ear to the demands of breakers that they begin monitoring channel nine at once. All they could do was lend their name and their reputation to the requests that a viable CB facility be established in this country by due process of law as soon as possible. In the face of extreme temptation and occasional vilification from the CB pirates REACT (UK) steadfastly maintained this awkward position right up to the bitter end. Despite their refusal to participate actively on the pirate band they were not altogether idle during the years, and by the time legalization became possible, then probable and then definite, a full-scale REACT monitoring service had been established nationally and was simply waiting for the big day. After the announcement in February 1981 that it was to be 27MHz FM by the autumn of the same year REACT decided it was time to come out of the closet and drop the supporters club bit; the time for action had arrived. The charter making REACT UK official and ready to roll was signed in the summer of 1981 and the full weight of the organization immediately became apparent. The relevant map explains it all rather more impressively and eloquently than I ever could.

What this means to you from now on should be immediately apparent. If it isn't – listen to me when I'm talking to you. Next time you're out in your car and it breaks down, or you become involved in or witness an accident you won't have to walk about in the middle of the night in a strange rural area looking for a telephone or any indication that the district has been inhabited during the past fifty years. All you will have to do is select channel nine and ask the nice REACT monitor for help. Just keep calm and answer all his questions. What could be simpler?

And if you ever see a bunch of thugs, or even one thug who appears to be a great deal larger than you are and may even be Luca Brazzi's big brother, administering a crash course in pain to someone who clearly doesn't deserve it – restrain your fighting instincts. Do not get out of your car and attack this giant bruiser, otherwise he will indubitably turn you into chutney. Just call up on channel nine and let the REACT monitor call the police; they're paid to be chutney from time to time.

In the light of this information you may, like others before you, feel that REACT is a noble cause worthy of your support. There are several ways you can help. The most obvious is to become a volunteer monitor; don't worry if you don't feel qualified for that because REACT will train you very well and by the time they've finished with you you'll know a whole lot more about CB than you ever will if you confine yourself to reading books on the subject and gossiping to your mates on the air.

If the commitment in terms of time is too great or you really don't think you could handle it then the next most obvious thing you can do for REACT is to send them money. Look on it like an AA subscription. You give them the cash, they buy the gear and train the monitors and then when you need help one dark and stormy night you know there'll be someone out there to answer you. It's just another kind of insurance really.

And if that sounds too expensive you can still do REACT, yourself and everybody else one last favour. Make sure that you know what to do in an emergency. Make sure you know what information the REACT monitor will require and that you are able to give it without delay. As a rule he (or she) needs to have a precise location, the nature of the problem which has arisen, whether anybody is injured and if so how many and how badly. He will also want you to remain on the scene until emergency services arrive. This latter is because he will want to stay in contact with the incident to ensure that help does arrive and if necessary to obtain further details of its location. Equally if not more important is the fact that in cases of personal injury the presence of a calm, uninjured person and the knowledge that help is on the way can sometimes make the difference between life and death. And if you could save someone's life just by talking to them you wouldn't hesitate, would you?

Remember all this. One day you might be the only base operator on the air to pick up a 10–33 and you really should know what to do about it. Better still, join REACT.

NATIONAL STRUCTURE PLAN OF REGIONAL ZONES, AREAS AND COUNTIES
(Refer to map)

REGIONAL ZONE CODE 001
SOUTH WEST ENGLAND
(Regional Director)

Area Zone Code 01 (area co-ordinator)
1 Gloucestershire/Avon
2 Wiltshire
3 Somerset
4 Dorset
   (Four county co-ordinators)

Area Zone Code 02 (area co-ordinator)
1 Devon
2 Cornwall
   (Two county co-ordinators)

REGIONAL ZONE CODE 002
SOUTH EAST ENGLAND
(Regional Director)

Area Zone Code 03 (area co-ordinator)
1 Oxfordshire
2 Buckinghamshire
3 Berkshire
4 Hampshire
   (Four county co-ordinators)

Area Zone Code 04 (area co-ordinator)
1 Surrey
2 West Sussex
3 East Sussex
4 Kent
   (Four county co-ordinators)

Area Zone Code 05 (area co-ordinator)
Greater London Area – area co-ordinator

Area Zone Code 06 (area co-ordinator)
1 Bedfordshire
2 Hertfordshire
3 Essex
   (Three county co-ordinators)

Area Zone Code 07 (area co-ordinator)
1 Cambridgeshire
2 Norfolk
3 Suffolk
   (Three county co-ordinators)

REGIONAL ZONE CODE 003
MIDLANDS
(Regional Director)

Area Zone Code 08 (area co-ordinator)
1 Leicestershire
2 Northamptonshire
   (Two county co-ordinators)

Area Zone Code 09 (area co-ordinator)
1 Staffordshire
2 West Midlands
3 Warwickshire
   (Three county co-ordinators)

Area Zone Code 10 (area co-ordinator)
1 Salop
2 Hereford and Worcestershire
   (Two county co-ordinators)

Area Zone Code 11 (area co-ordinator)
1 Derbyshire
2 Nottinghamshire
3 Lincolnshire
   (Three county co-ordinators)

When you need help on the highway...call REACT on citizens radio emergency channel 9...

Citizens Two-Way Radio is a low-cost, convenient means of providing two-way communications from your automobile to home or business. It is as simple to operate as a TV set and easier than telephone. No tests or special technical knowledge is required. Any U.S. resident over 18 years of age may apply and obtain a FCC license to operate Citizens Radio. One out of 10 automobiles is already equipped with Citizens Two-Way Radio

REACT is a nation-wide organization of over 1,500 volunteer group totaling approximately 100,000 volunteers who utilize equipment in the Citizens Radio Service to monitor Emergency Channel 9 and provide local two-way radio communication in response to emergencies.

REACT teams are prepared to provide supplementary communications in any emergency. Effective local 2-way radio communications has proved valuable whenever normal telephone communications is interrupted because of fire, blizzard, earthquake, flood, hurricane, tornado, or other disasters.

Through a cooperative understanding between the American National Red Cross and REACT, local teams are encouraged to participate in their community's pre-disaster

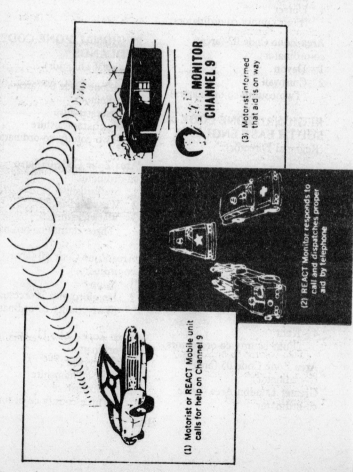

MONITOR CHANNEL 9

(3) Motorist informed that aid is on way

(2) REACT Monitor responds to call and dispatches proper aid by telephone

(1) Motorist or REACT Mobile unit calls for help on Channel 9

REGIONAL ZONE CODE 004
WALES

Area Zone Code 12 (area co-ordinator)
1 Clwyd
2 Gwynedd
   (Two county co-ordinators)

Area Zone Code 13 (area co-ordinator)
1 Powys
2 Dyfed
3 West Glamorgan
4 Mid Glamorgan
5 Gwent
6 South Glamorgan
   (Six county co-ordinators)

REGIONAL ZONE CODE 005
NORTHERN
(Regional Director)

Area Zone Code 14 (area co-ordinator)
1 Greater Manchester
2 Merseyside
3 Cheshire
   (Three county co-ordinators)

Area Zone Code 15 (area co-ordinator)
1 North Yorkshire
2 West Yorkshire
3 South Yorkshire
4 Humberside
   (Four county co-ordinators)

Area Zone Code 16 (area co-ordinator)
1 Cumbria
2 Lancashire
   (Two county co-ordinators)

Area Zone Code 17 (area co-ordinator)
1 Northumberland

2 Tyne and Wear
3 Durham
4 Cleveland
   (Four county co-ordinators)

REGIONAL ZONE CODE 006
SCOTLAND
(Regional Director)

Area Zone Code 18 (area co-ordinator)
1 Lothian
2 Borders
   (Two county co-ordinators)

Area Zone Code 19 (area co-ordinator)
1 Strathclyde
2 Central
3 Dumfries and Galloway
   (Three county co-ordinators)

Area Zone Code 20 (area co-ordinator)
1 Highland
2 Grampian
3 Tayside
   (Three county co-ordinators)

REGIONAL ZONE CODE 007
NORTHERN IRELAND
(Regional Director)

Area Zone Code 21 (area co-ordinator)
1 Antrim
2 Tyrone
3 Fermanagh
4 Armagh
5 Down
   (Five county co-ordinators)

**Regional Zone 008 –
   Isle of Man**
**Regional Zone 009 –
   Channel Isles**

REACT Serves you on Citizens Radio Emergency Channel 9

The U.S. Federal Communications Commission and the Canadian Dept. of Communications have reserved Channel 9 for emergency messages and motorists assistance. Your neighbors who are members of the local REACT team are volunteer Channel 9 monitors. They strive to improve traffic safety by using Citizens Radio Channel 9 to:

- Report Accidents • Summon medical aid faster
- Keep traffic moving • Report road conditions
- Give road directions • Avoid being lost

REACT Objectives:

1. To assist in all forms of local emergencies by furnishing instant radio telephone communications in cooperation with proper authorities and official agencies.
2. To maintain and encourage operating efficiency through proper communication techniques.
3. To operate and maintain equipment at peak efficiency and in accordance with F.C.C. regulations.
4. To promote the proper and effective use of the official CB Emergency Channel 9.

All users of Citizens Two-Way Radio are requested to cooperate toward the success of the Official Emergency Channel. A successful emergency network will add greatly to the value of your radio equipment. You can help!

1. Confine communications on Channel 9 to emergencies and motorists assistance, in accordance with FCC regulations.
2. Allow qualified monitors to answer emergency calls first. If no REACT monitor or other organized monitor responds, then respond as an individual.
3. If you are interested in joining this public service movement, contact your local REACT team. If there is no team in your community, contact REACT National Headquarters for information on how to form a team.

Our objective is to eventually provide sufficient coverage so that you can call a REACT monitor at any time, anywhere and get assistance on Channel 9.

REACT International, Inc.
111 E. Wacker Drive, Chicago, IL 60601

EXECUTIVE BOARD OF DIRECTORS

Chief Executive Director
(Chairman REACT-UK)
Mr A. A. Joiner, JP
3 Bridge Cotterell
Bristol Avon

Deputy Chief Executive Director
(Vice-chairman REACT-UK)
Mr D. Wright, BSc
11 Tewksbury Road
St Werburghs
Bristol
Avon

Executive Managing Director
(REACT-UK co-ordinator)
Mr A. Mackay
Strathnaver
142 Lutterworth Road
Nuneaton
Warwickshire
CV11 6PE
Tel: (0203) 383005

Executive Director
(team registration)
Mr P. D. Horne

10 Buckingham's Way
Sharnford
Leicestershire
LX10 3PX

Executive Director
(administration – finance)
Mrs J. E. Mackay, BSc
Strathnaver
142 Lutterworth Road
Nuneaton
Warwickshire

Executive Director
(REACTer UK editor)
Mr V. Bull
17 Church Road
Harold Park
Romford
Essex
RM3 0JX

Executive Director
Insp. J. B. Cambell
Devon and Cornwall
Constabulary H.Q.
Middlamoor
Exeter
Devon

A large number of breakers in the early pirate days seemed to be increasingly fascinated with the technology of the equipment they were using. This is a reflection of what happened in America after CB became popular there; not long after the boom in sales of CB sets there was a marked rise in the number of applications for all types of radio licences, as more and more people found that radio was far more useful than they'd realized and were now looking for a high-grade service, or found that it was simply far more interesting than they had realized and now wished to move on to bigger and better things.

It is almost certain that a similar pattern will emerge in this country as well; indeed it has already begun. There are more than a few pirates on the AM band who are far more interested in the possibilities of working skip than they are in getting rescued when they run out of petrol. More often than not they will be running fairly powerful burners, operating on Sideband and may possibly even have paid a visit to the rig doctor and provided themselves with a stunning array of outband channels where they can DX undisturbed.

All of this is well outside the terms of reference for a CB facility, which is by definition supposed to provide publicly accessible two-way radio communication over short distances.

Most of these DX pirates would probably have a great deal more fun, and more to the point a great deal more success, if they were to leave the CB bands to their designated usage and move onto the Amateur frequencies. It's quite likely that many of them would agree; it's even probable that many of them have already done so. This is not necessarily a situation which finds favour with the pirates, the Hams or the Home Office, for a variety of reasons.

In the main the pirates' objections to this are likely to stem from two sources. The first is that, because of its more sophisticated nature and greater capabilities, proper Ham equipment is vastly more expensive than CB units. It has to be, because its potential for causing interference is greater, mostly because of its higher power, and therefore manufacturing standards have to be much more rigorous. This costs money. If you like, it's the difference between a Roller and a Mini. Secondly you can't just walk into a Post Office, cough up your eight quid and walk out with an Amateur licence. You have to pass exams. If you're going to be allowed to sweep the world with your radio transmissions this isn't an altogether unreasonable attitude. It would be nice if you knew what you were doing before you started. So before a licence is issued candidates, who must be over 14, must pass two exams. One is concerned with Licensing Conditions and Interference and the other with Operating Practices, Procedures and Theory. Between them the two papers will take up almost three hours of your time, and examinations are conducted by the City and Guilds of London Institute, a body long established in the field of professional and craft qualifications, so they're by no means a formality.

Recent advances in technology have made the type of equipment available to radio Hams far simpler to use and maintain than previously. In fact black box technology has invaded even this hallowed sanctum, which has caused severe rifts between existing licensed Hams already. Many of the pirates who have worked skip on 27MHz feel that most of the Ham gear is no more difficult to use than a Sideband CB rig, and fail to see the necessity for passing exams. This, coupled with the extra costs, has meant that a vast horde of radio enthusiasts who should be working the Amateur bands are in fact still haunting the CB network and using it for their own purposes.

The situation gets worse still. The simplest and cheapest way for a newly licensed Ham to get on the air is on the 2-metre band. Base and mobile equipment for this service is only marginally different to that used on 27MHz CB; it is simplicity itself to operate, just like a rig. And just like many CB users, a great number of Hams use their mobile 2-metre sets as base units.

Anyone with an average knowledge of CB, particularly if they are used to skip talking on Sideband, where the jargon and codes are almost exclusively based on the established Ham format, would find little difference if they switched to 2 metres. Equipment, procedures and technology are virtually identical. Two-metre equipment responds as well to the judicious application of a

If you were a Ham this is the sort of thing you'd find sticking up on a pole outside your house. Just a fairly straightforward groundplane antenna, clean and simple.

170

burner as does 27MHz. All of which probably explains why so many DX pirates have moved on to 2 metres. The pity is that hardly any of them have bothered to get the legal feeling and take the exams. In the main they seem to be so used to being pirates that the prospect no longer bothers them. In all probability they're less likely to suffer prosecution on 2 metres than they ever were on 27MHz.

While it is difficult to condone their attitudes in this matter it is no easier to vilify them, particularly once you have accepted the principles inherent in a CB service – free access to the airwaves for anyone who can press buttons effectively. However, the law, while it is arguably an ass in more areas than this one, remains the law. And while we may be tempted to suspect that the attitude of authority towards public use of the airwaves may well undergo subtle changes as a result of the introduction of a successful CB facility it would be naïve in the extreme to imagine that any such change would be likely to manifest itself at any noticeable speed.

Whatever your opinions on all this you should know that Ham radio is a jolly good thing as far as anyone who develops a more than passing interest in the technology of radio is concerned. Whether such interest is generated in consequence of legal CB use or wholescale DX piracy appears to be of very little importance. Amateur band is the logical next step if you enjoy radio for its own sake.

You can find out more about it from the Home Office Radio Regulatory Department, Waterloo Bridge House, London SE1, who are the people in charge of this bit of airspace as well as the bit you're used to. Passing your first exam will give you a Class B licence, which will give you legal access to 2 metres, and further study, involving word-perfect proficiency in Morse at the rate of 12 words per minute, will allow you to collect your Class A licence. Armed with this you may, if you have the desire, the time and the money, operate on up to 23 wavebands with a maximum power, in some cases, of 400 Watts. Makes your 40-channel, 2-Watt ERP CB rig look small, does it not?

A full set of rules, syllabus and objectives for the exams can be obtained from the City and Guilds of London Institute, Electrical and Telecommunications Branch, 76 Portland Place, London W1N 4AA. Send them 80p and ask for pamphlet number 765, RAE. The letters stand for Radio Amateur's Examination, which is what you want to take.

If you don't think you're particularly good at examinations then you ought to know about the Radio Society of Great Britain (RSGB). These nice people live at 35 Doughty Street, London WC1 2AE, and can help you a great deal. They have about 23,000 members and principally act as an organizing body and spokespeople for Amateurs. For beginners they offer a range of publications and services, and perhaps the most useful will be their guides to Amateur Radio in general or to the examination itself, together with the Radio Communication Handbook. They also broadcast very slow Morse transmissions for beginners, which can be a big help.

Once you've passed, the list of options is considerably larger than it ever will (or could) be for CB. You may send speech, Morse, teletype, even TV signals as an Amateur, and beam off around the world. It's quite possible, if you're a gadget freak – and you almost certainly are if you've gone this far – that you'll soon have one of these little home computers to play with. You may or may not already be aware that computers are stupid. Their comprehension is limited to yes or no, and their skill comes solely from being able to ask a question and assimilate the yes or no answer millions of times a second. All of which makes speaking to computers dead boring. However, they can do one smart thing, and that's speak to each other. So if your computer knows something your brother's doesn't – they can tell each other. Just plug them in and away they go. Which is great if you live next door, not so funny if you're miles apart.

This little problem can be overcome by using the telephone and an audio coupler, but we already know that phone calls cost money. It can also be overcome by using radio; high-speed transmission of a series of bleeps which computers understand. This is clearly much cheaper. It is less private, of course, but unless you're Oleg Penkovsky that shouldn't worry you. Using CB for this purpose has a number of attractions, which is probably why so many people are at it in the States. Sadly the licence for CB in this country allows only the sending of signals in plain speech except for selective calling and digital transmitter identification. Which means data transmission between computers is not allowed. In any case the low range of CB makes it less attractive than Amateur Band for this purpose.

172

The future of personal communications of this nature is also less suited to CB than anything else; CB is a convenient peg to hang it all on, but in the main it's only the focus for an ideal, the identification tag of a whole new principle.

Cordless phones are a beginning. Basically they're only small, low-powered hand-helds which allow you to use your telephone to make or receive calls anywhere you like within a radius of, say, one mile from the base installation at your home or office. The same principle is applicable to CB; mobile sets frequently appear as one-handers, with all the command functions plus the loud-speaker on the microphone, which is connected to the set by one cord. Silicon chip technology allows us to replace that cord with radio waves, which means that you will be able to monitor and use your CB provided you are within about a mile of your car.

Current thinking in America points to the eventual combina-

Mobile Ham equipment below a car dash. Not terribly different to CB, except that the frequency in use is infinitely variable by minute degrees rather than in fixed channels.

tion of these advances with the already-extant phone-patch, which allows CB/phone link-up, leading eventually to the point where you will be able to make or receive direct-dialled calls from the other side of the world as long as you are within a mile of your house, car or office, probably on your wristwatch. Beam me aboard, Scotty. The distinction between radio and telephone networks is already fuzzy; it's likely that it will soon disappear. Likely also that it will be able to incorporate all the other microchip miracles associated with computers, calculators and everything else. We are already capable of making TV screens very small indeed and we have just begun to dispense with what was the last remaining bulk object in a TV – the tube. Once that's been scrapped we'll be talking about video-phones. And since there will hardly be any discernible difference between phones and radio, we'll be dealing with video-radio; talking transatlantic pictures in your wristwatch.

Just think; all you ever wanted was to be able to order your hamburger before you arrived, and now here you are trying to come to terms with The Prime Directive. But don't worry, because it won't happen that quickly. We've got the technology all right,

Just like the mobile unit in every respect – because it is the mobile unit. Immensely versatile, this one can be unplugged from its dashboard location, hung on a shoulder strap and carried anywhere. Clever stuff.

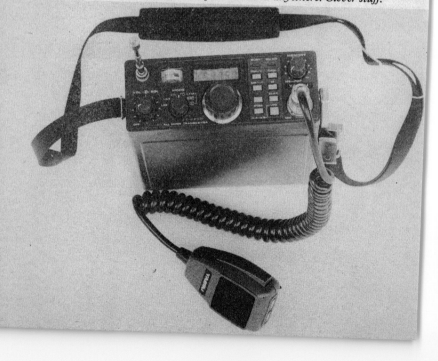

but we'll never learn to cope with the bureaucracy. Look at the fuss involved with the introduction of CB. Can you imagine us trying to handle the economics of direct-dial video radio calls between two unidentified wristwatches in Leeds, never mind if one of them happens to be in New York.

You know how to speak with the dead? Dial 100 and listen; the silence speaks for itself. And it's not the fault of British Telecom, *per se*. It's just that their staff cannot cope with their existing technology at an operational level. The signals are willing, but the flesh is weak. The trouble is that progress is all so fast that it's leaving the average punter behind. We've got the technology, yes. It's just that we don't know what to do with it. Next time a gadget goes wrong – calculator, computer, telephone – don't kick it. If you call up Swagman on your CB and get put through to the speaking clock it probably wasn't the fault of the radio at all; more likely it was some poor mystified human being trying to assimilate microfiche and chips and failing miserably. . . .

11 BUYING A RIG

It seems more than likely that CB will be the only growth industry of the eighties in this country. Perhaps that's unfair, but still.

When CB first came to these shores in any quantity and with any real degree of understanding, it came in the bottom of suit-cases, in spare-tyre wells, shoved under car seats. In short, it was smuggled here in ones and twos. By the end of the seventies it was still being smuggled, but in bulk. CB rigs were widely available.

In the main they could be bought in pubs and, latterly, from the host of shops which sprang up quite legally selling the accessories which make CB work. The fact was that although transceivers were the subject of import control and bans the other things – linear amps, antennas, power mikes – were not. It was quite legal to sell a customer everything he needed to make a CB rig work except the unit itself.

There is no doubt that the CB shops were selling rigs though, and by 1980, maybe 1981, more and more openly. In most cases the shops were small single-fronted affairs, probably a closed-down sweetshop or similar. The majority had no window displays at all; some were lucky to have a counter or a till. They were mainly dead scruffy.

But apart from CB accessories they all dispensed tea, advice, natter and goodwill indiscriminately. They were nice places to go shopping, almost Victorian in the way all their callers received such personal attention; as long as the piracy continued the shops too were part of the club. In some cases they *were* the club.

With the advent of the legal FM system that whole aspect of CB is more or less doomed; the writing may not actually be on the wall, but it is writ large in MPT 1320.

It was long ago estimated by the National Electronics Council that the CB market in this country would be somewhere equiv-alent to the domestic hi-fi level; measured in hundreds of millions of pounds. In other words very big business indeed. It would be naïve to assume that the retail giants would ignore such a money-spinner. They didn't.

176

Well before legalization a great number of large multiples – including electrically orientated firms like Dixon's and others not so well aligned, like W. H. Smith – were making plans for CB in-store promotions for two or three weeks at a time simultaneously in 250 retail outlets across the nation. The little shops, despite their friendly atmosphere, couldn't compete. Like the supermarket was to the corner grocer, the multiples were the Goliath which would stamp David into oblivion. The buying habits of the CB world were changing. And changing to suit the needs of big business, not its customers.

Somewhere in the retail jungle treads a punter; all he needs is a CB. Armed with the desire and a Barclaycard he must sally forth to meet and do battle with the salesmen of the world. Basically he's got no chance. Despite the training schemes most of the salesmen will know as much about CB as the captain of the *Titanic* knew about icebergs. At least to begin with.

Before he goes, the punter might like to know what's on offer. In the world of the microchip almost anything will be obsolete before it's reached the shops, and any kind of guide will be of very little practical value. Nevertheless here is a guide to what was already available *before* legalization. No doubt there are more and better units around by now, which can do far more varied and wonderful things than ever before. This used to be called planned obsolescence, and is generally considered to be very good business, but be careful; bank managers call it profit or loss, depending on whether you're buying or selling.

AMSTRAD

Two in-car units are available, although base units will follow later. The most noticeable factor about both CB900 and CB901 is their size. They are the smallest sets so far available, and only 1½ × 5½ inches. Both use LED indicators extensively for power and signal strength monitoring and transmit/receive indicators. CB900 should sell for less than £70* while CB901, which has everything the 900 has plus Roger Beep and automatic squelch will be about £85. Both have integral speaker and come boxed with their fitting kits and instructions.

*Note: all prices are correct at the time of going to press but they are inevitably liable to changes beyond the author's control.

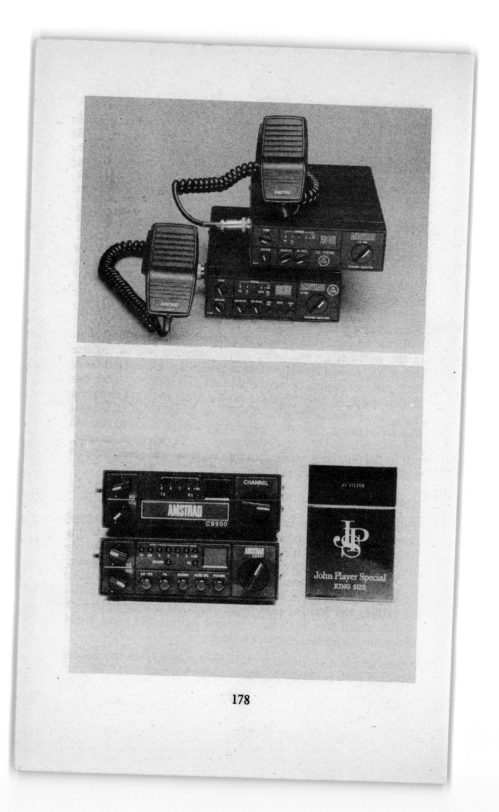

BINATONE

This is a fairly extensive range covering most available options for CB users. The basic 40-channel mobile rig, Speedway, is just that – basic. A minimum of controls and features will perhaps make it one of the cheaper 'starter' units in the shops. The Route 66 model is more complex, with transmit and receive indicators and a PA facility, although still not overcrowded with gadgets.

More interesting is the 5 Star, with RF gain, mike gain, delta tune, PA, automatic channel 9 monitoring and a distant/local switch concealed behind the title of hi/lo TX power. It also has switchable internal/external speaker options. The faceplate looks busier than the other two, as indeed it is.

Not yet available is the Longranger base unit which has exactly the same features as the 5 Star plus a headphone socket. This may seem like another gadget, but for home use – not annoying your family – or especially when speaking with faint stations, it should be very useful.

Two handsets are also available. The basic 3-channel unit puts out its maximum allowable four watts from the transmitter, has distant/local option (again called hi/lo), contains rechargeable batteries and has facility for external speaker, antenna and mains adapted power source. The 12-channel does all of this as well, only on more channels, and also features a combined power and S-meter which also shows battery condition. Both units have a built-in full-length antenna as well.

And on the subject of antenna length, there are also two SWR meters as well, reasonably similar except that Model 02-5771 can deal with 200 Watts as opposed to the 100 of which 02-5798 is capable.

FIDELITY

Two mobile units which come as a basic 40-channel rig with volume, squelch and S-meter, the CB1000FM, or the CB2000FM which includes channel 9 priority, RF gain and mike gain, external speaker option and a dimmer switch for the LED channel indicator. Both units come with fitting kit and instructions and are the forerunners of a somewhat more extensive range.

As a matter of interest both have the 10db attenuator, switchable at the rear of the unit should you wish to double up as a base unit using an antenna fixed at more than 23 feet from the ground.

The basic unit should sell for about £60, while the CB2000FM is likely to cost closer to £80.

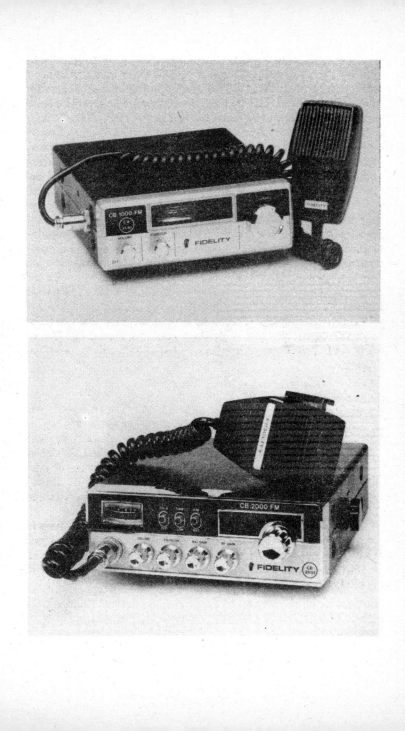

GRANDSTAND

Two mobile units again, although interestingly even the basic unit, SY, features RF gain, a noise blanker and channel 9 priority.

The 330 mobile unit also has indicators for transmit and receive, signal and power meter and a clarifier.

The 530 base unit has all that plus tone control, built-in SWR meter, dimmers for the channel display and a headphone socket.

No prices are yet available for this attractive-looking range, or for the extras – preamp desk mike, meters, testers, etc. – which are due soon. An extensive use of LED and coloured lighting – the linear display for signal, RF power or PA power on the 330 mobile, the digital clock in the 530 base set – are probably going to catch a number of eyes, though, and the importers expect the sets to be priced comparably with everybody else.

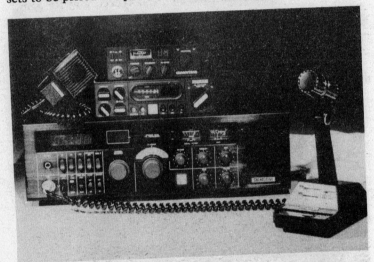

HAM INTERNATIONAL

One of the best-known and most respected names in CB already, Ham are famous for their extensive and superb range of AM equipment, all the way up to the recent 120-channel Concorde II mobile unit, offering 120 channels on AM, FM and upper and lower sideband.

To meet the specifications of MPT 1320 Ham are producing two mobile units, the basic Explorer and the more complex Mariner which features Roger Beep, standby, noise blanker, tone and RF gain and just about everything else. Explorer should cost about £75 and Mariner about £110, while the Hercules base unit, at £165, seems to have every conceivable gadget available plus warning lights for everything in sight.

INTERCEPTOR

A brand-new name for CB, and a whole list of brand-new features into the bargain. These sets (five are planned) are not going to be cheap – they start at £80 – but they are going to be good, using purpose-designed chips and circuitry.

The basic 80 quid mobile has all you'd expect, plus RF gain, and signal and power metering is by progressive coloured light readout, going up from green through amber to red. Also featured is SWR protection (it switches off above 25:1) and reversed polarity protection.

Up one step to £100 and you get push-button channel select, mike gain, a digital clock and all the pretty lights.

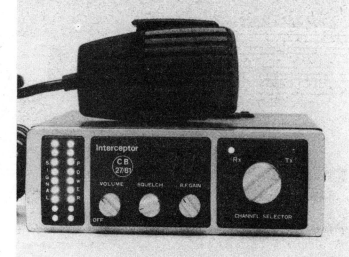

A big first for British CB is the £125 one-hander, with channel change, volume and squelch on the handset. New departure is that as well as the power pack, which you can install under the bonnet or in the boot, the unit also includes a dash pack with LED readouts for channel in use, the coloured metering lights and a digital clock.

For £200 or thereabouts the first CB/radio/cassette will be available. All the CB goodies in the range in one unit with your radio and stereo.

A home base is also coming, for about £180, with channel scanner, all the controls – RF gain, mike gain, clarifier, etc. – you need. Any of these not operated by the slider principle will be worked by a digital keyboard. Should be very nice, and all British.

JOHN WOOLFE RACING

Best known for importing performance parts for American cars, JWR have stepped into CB in a big way and are bringing in three of their own branded units, M1, M2 and M3.

Strangely, M2 is the basic 40-channel mobile for about £65, with S-meter/RF meter, PA facility and transmit indicator light. Interesting features on this include dual polarity, a 10db attenuator and built-in SWR protection which can stand transmitting at 20:1 for up to five minutes.

M1, which is about £85, has distant/local, channel 9 priority, and channel free or busy indicator lights on top of all that, while M3, at about £100 adds RF gain, noise blanking, PA and transmit/receive indicators, but dispenses with the channel free/busy lamps.

MAJOR

An established name in Europe, so far only details on one basic mobile unit have been released as a foretaste of a fuller range to follow shortly. The 3000 is basic enough, but still features PA, RF gain and delta tune, plus built-in SWR protection for up to five minutes.

Their AM sets are very popular indeed abroad and if they manage to introduce a comparable range built to MPT 1320 then it's likely that the sets will sell very well and be much appreciated by the users.

MIDLAND

Yet another well-known name fresh from AM transceivers, with a strong reputation for quality and reliability. A wide range of equipment to MPT 1320 has been announced, beginning with the 75-720 three-channel handset, very basic at about £60. Far more interesting is the 40-channel handset 77-810 at just over £70. Still basic, but 40 channels . . .

Basic mobile unit is the 2001, which at £70 offers PA, power meter, external speaker facility and transmit indicator light, plus five minutes SWR protection built-in.

The 3001 is about £80 and the extra gets you distant/local, RF gain and switchable low pass filter on top.

Going up to £90 will buy the 4001 which provides a selection of indicator lights and mike gain as well. All the range advances in easy stages, and each stage adds features without losing or swapping like other manufacturers, which makes it easy to follow and see exactly where your money goes.

There is also a base unit, 79-200, which, at £150, is reasonably equipped with hi/lo RF power, external speaker, S/RF meter and the normal controls.

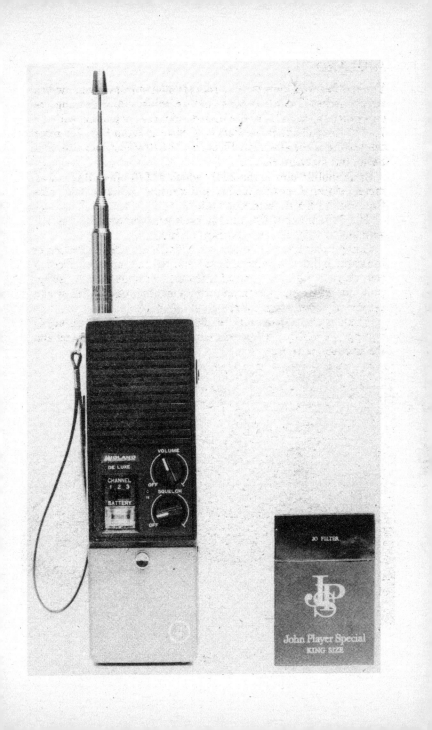

RADIOMOBILE

Long associated with in-car audio products, as the name implies, Radiomobile have introduced two mobile units, CB201 and CB202.

201 is the basic 40-channel unit with LED displays for signal and power metering, while 202 also features RF gain, channel 9 priority and a PA facility. No prices yet available, but no prizes for guessing which is the more expensive of the two units.

Like most other British companies Radiomobile are taking CB very seriously and see it as a major growth market for at least the next five years. Therefore expect to see further units and accessories fairly soon.

SULKIN

This range will be marketed under the York brand name, which is already established in the field of electronics and audio equipment and includes a complete range of rigs and accessories including SWR meters. They are also the first British manufacturer to produce a complementary range of antennas, both conventionally mounted and magnetic or clamp-mounted, plus extension and PA speakers.

There are two rigs – a basic 'cooker' for about £70 and at £90 a more comprehensive item with RF gain, mike gain, channel 9 priority and tone control.

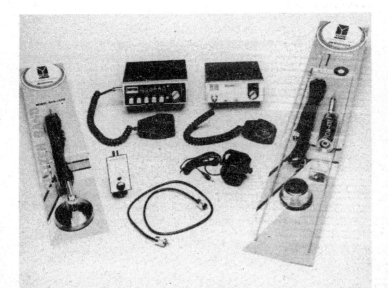

12 THE BRITISH SPECIFICATIONS – MPT 1320 AND MPT 1321

Foreword

1   Citizen's band radio, a personal two-way radio system, is available for use throughout the United Kingdom. It operates in the 27 MHz and the 934 MHz bands.

2   The Wireless Telegraphy Act 1949 provides that no radio equipment may be installed or used except under the authority of a licence granted by the Secretary of State. All citizens band radio equipment, whether hand held, mobile or base station, must be covered by a licence; it is a condition of this that the apparatus fulfils, and is maintained to, certain minimum technical standards.

3   The manufacturer, assembler, or importer of citizens band equipment is responsible for ensuring that the apparatus conforms with the specification; and any additional requirements imposed by regulations under the Wireless Telegraphy Act 1949. Conformity with the required standards may be established by tests carried out by the manufacturer, assembler or importer, or by a reputable test establishment acting on his behalf, but in either case conformity with the specification will remain the responsibility of the manufacturer, assembler or importer.

MPT 1320

1 General

1.1 Scope of specification
This specification covers the minimum performance requirements for angle modulated radio equipments, comprising base station, mobile and hand held transmitters and receivers or receivers only and additionally any accessories, for example attenuators, vehicle adaptors for optional use with the above for use in the Citizens Band Radio Service.
For all equipments covered by this specification, the nominal separation between adjacent channel carrier frequencies is 10 kHz.

1.2 Permitted effective radiated power
The output radio frequency power of the equipment is limited to 4W. With the antenna permitted by the conditions of the licence for use with the equipment this provides an effective radiated power of 2W. (*See note.*)
If an antenna is mounted at a height exceeding 7m the licence will require a reduction in transmitter power of 10 dB.
To enable the user to accomplish this easily, the equipment manufacturer shall provide as a standard facility on the equipment means by which the transmitter output power may be reduced by a minimum of 10 dB.

1.3 Operating frequencies
The equipment shall provide for transmission and reception only of angle modulated emissions on one or more of the following radio frequency channels:

Channel	1	27.60125 MHz	Channel	14	27.73125 MHz
,,	2	27.61125 ,,	,,	15	27.74125 ,,
,,	3	27.62125 ,,	,,	16	27.75125 ,,
,,	4	27.63125 ,,	,,	17	27.76125 ,,
,,	5	27.64125 ,,	,,	18	27.77125 ,,
,,	6	27.65125 ,,	,,	19	27.78125 ,,
,,	7	27.66125 ,,	,,	20	27.79125 ,,
,,	8	27.67125 ,,	,,	21	27.80125 ,,
,,	9	27.68125 ,,	,,	22	27.81125 ,,
,,	10	27.69125 ,,	,,	23	27.82125 ,,
,,	11	27.70125 ,,	,,	24	27.83125 ,,
,,	12	27.71125 ,,	,,	25	27.84125 ,,
,,	13	27.72125 ,,	,,	26	27.85125 ,,

Note: The licence requires that equipments which have provision for the connection of an external antenna shall not be connected to other than a single element rod or wire antenna not exceeding 1.5m in overall length.

191

Channel 27	27.86125 MHz		Channel 34	27.93125 MHz
" 28	27.87125 "		" 35	27.94125 "
" 29	27.88125 "		" 36	27.95125 "
" 30	27.89125 "		" 37	27.96125 "
" 31	27.90125 "		" 38	27.97125 "
" 32	27.91125 "		" 39	27.98125 "
" 33	27.92125 "		" 40	27.99125 "

Citizens band radio equipment shall not contain facilities for transmission of radio frequencies other than those listed above, and those contained in MPT 1321.

Single channel equipment may be tested on any one of the approved channels. Multi-channel equipment shall be equipped to operate at the centre, and the upper and lower limits of the frequency range over which channel switching is possible.

1.4 Permitted modulation

Only equipment which employs angle modulation and has no facilities for any other form of modulation will meet the requirements of this specification.

1.5 Labelling

The equipment shall be provided with a clear indication of the type number and name of the manufacturer.

1.6 Certification of compliance

Compliance with this specification shall be indicated by a mark stamped or engraved on the front panel of the equipment.
The mark used to indicate compliance shall be as shown in Fig. 1.

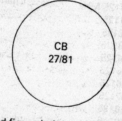

CB
27/81

Letter and figure height not less than 2mm.

Fig. 1.

1.7 Controls

Those controls, which if maladjusted might increase the interfering potentialities of the equipment, shall not be easily accessible.

2 Test conditions: Atmospheric conditions and power supplies

2.1   General

Tests shall be made under normal test conditions (Clause 2.3) and also, where stated, under extreme test conditions (Clause 2.4).

2.2   Test power source

During tests, the power supply for the equipment may be replaced by a test power source, capable of producing normal and extreme test voltages as specified in Clauses 2.3.2 and 2.4.2.

The internal impedance of the test power source shall be low enough for its effects on the test results to be negligible.

For the purpose of tests, the supply voltage shall be measured at the input terminals of the equipment.

If the equipment is provided with a permanently connected power cable, the test voltage shall be measured at the point of connection of the power cable to the equipment.

During the tests of the power source voltage shall be maintained within a tolerance of $\pm3\%$ relative to the voltage at the beginning of each test.

In equipment in which batteries are incorporated, the test power source shall be applied as close to the battery terminals as practicable.

2.3   Normal test conditions

2.3.1   *Normal temperature and humidity*

The normal temperature and humidity conditions for tests shall be any convenient combination of temperature and humidity within the following ranges:

   Temperature 15°C to 35°C

   Relative humidity 20% to 75%

When it is impracticable to carry out the tests under the conditions stated above, a note to this effect stating the actual temperature and relative humidity during the tests, shall be added to the test report.

2.3.2   *Normal test source voltage*

2.3.2.1   Mains voltage

The normal test source voltage for equipment to be connected to the mains shall be the nominal mains voltage. For the purpose of this specification, the nominal voltage shall be the declared voltage or any of the declared voltages for which the equipment was designed. The frequency of the test power source corresponding to the AC mains shall be between 49 and 51 Hz.

2.3.2.2   Regulated lead-acid battery power sources

When the radio equipment is intended for operation from the usual type of regulated lead-acid battery source, the normal test source voltage shall be

1.1 times the nominal voltage of the battery (6 volts, 12 volts, etc).

2.3.2.3   Other power sources

For operation from other power sources or types of battery, either primary or secondary, the normal test source voltage shall be that declared by the equipment manufacturer.

2.4   Extreme test conditions

2.4.1   *Extreme temperatures*

For tests at extreme temperatures, measurements shall be made in accordance with the procedures specified in Clause 2.5 at an upper value of +45°C and at a lower value of −5°C.

2.4.2   *Extreme test source voltages*

2.4.2.1   Mains voltage

The extreme test source voltages for equipment to be connected to an AC mains source shall be the nominal mains voltage ±10%. The frequency of the test power source shall be between 49 and 51 Hz.

2.4.2.2   Regulated lead-acid battery power sources

When the equipment is intended for operation from the usual type of regulated lead-acid power source, the extreme test voltages shall be 1.3 and 0.9 times the nominal voltage of the battery.

2.4.2.3   Other power sources

The lower extreme test voltage for equipment with power sources using primary batteries shall be as follows:

a.   For Leclanche type of battery −0.85 times the nominal voltage
b.   For mercury type of battery −0.9 times the nominal voltage
c.   For other types of primary battery − end point voltage declared by the equipment manufacturer.

For equipment using other power sources, or capable of being operated from a variety of power sources, the extreme test voltages shall be those declared by the equipment manufacturer and shall be recorded with the test results.

2.5   Procedure for tests at extreme temperatures

2.5.1   *General*

Before making measurements, the equipment shall be placed in a temperature controlled chamber for a period of one hour or so for such period as may be judged necessary for thermal balance to be obtained. The equipment shall be switched off during the temperature stabilisation period. The sequence of tests shall be chosen and the humidity content in the test chamber shall be controlled so that excessive condensation does not occur.

2.5.2   *Test procedure*

For tests at the upper temperature, after thermal balance has been attained (Clause 2.5.1), the equipment shall be switched on for 1 minute in the transmit condition followed by 4 minutes in the receive condition, after which the appropriate tests shall be carried out.

For tests at the lower temperature, after thermal balance has been attained (Clause 2.5.1), the equipment shall be switched on for 1 minute in the receive condition after which the appropriate tests shall be carried out.

3 Electrical test conditions

3.1 Transmitter artificial load

Tests on the transmitter shall be carried out using a 50 ohm non-reactive, non-radiating load connected to the antenna terminals. If necessary an impedance matching device may be used for testing.

3.2 Test fixture
3.2.1 *General*

A test fixture will be required to permit relative measurements to be made on the sample.*

This test fixture shall preferably provide a 50 ohm radio frequency terminal at the working frequencies of the equipment.

The test fixture shall provide input and output audio coupling and a means of connecting an external power supply.

The following characteristics shall apply to the test fixture:

a. The coupling loss shall be as low as possible, and in any case not greater than 30 dB;

b. The variation of coupling loss with frequency shall not cause errors in measurement exceeding 2 dB;

c. The coupling device shall not incorporate any non-linear elements.

3.3 Test site and general arrangements for measurements involving the use of radiated fields.
3.3.1 *Test site*

The test site shall be located on a surface or ground which is reasonably level. At one point of the site, a ground plane of at least 5 metres diameter shall be provided. In the middle of this ground plane, a non-conducting support, capable of rotation through 360° in the horizontal plane, shall be used to support the test sample at 1.5 metres above the ground plane.

The test site shall be large enough to allow the erection of a measuring or transmitting antenna at a distance from the test sample of not less than half the wavelength corresponding to the lower frequency to be considered. The distance actually used shall be recorded with the results of the tests carried out on the site.

Sufficient precautions shall be taken to ensure that reflections from extraneous objects adjacent to the site, and ground reflections do not degrade the measurements.

---

*Note: Any connections provided on the equipment in order to facilitate relative measurements, shall not affect the performance of the equipment either in the test fixture or when making measurements involving the use of radiated fields.

3.3.2 *Test antenna*

The test antenna is used to detect the radiation from both the test sample and the substitution antenna, when the site is used for radiation measurements. This antenna is mounted on a support capable of allowing the antenna to be used either horizontally or vertically polarized and for the height of its centre above ground to be varied over the range 1–5 metres. Preferably test antenna with pronounced directivity should be used. The size of the test antenna along the measurement axis shall not exceed 20% of the measuring distance. For radiation measurements, the test antenna is connected to a test receiver, capable of being tuned to any frequency under investigation and to measure accurately the relative levels of signals at its input.

3.3.3 *Substitution antenna*

The substitution antenna shall be a $\lambda/2$ dipole resonant at the frequency under consideration, or a shortened dipole, calibrated against the $\lambda/2$ dipole. The centre of this antenna shall coincide with the reference point of the test sample it has replaced. This reference point shall be the point at which the external antenna is connected.

The distance between the lower extremity of the dipole and the ground shall be at least 0.3m.

The substitution antenna shall be connected to a calibrated signal generator when the site is used for radiation measurements.

The signal generator and the receiver shall be operating at the frequencies under investigation and shall be connected through suitable matching and balancing networks.

3.4 Normal test modulation

Where stated, the transmitter shall have normal test modulation as follows: The modulation frequency shall be 1 kHz and the resulting frequency deviation shall be 60% of the maximum permissible frequency deviation (Clause 4.3.1).

4 Transmitter

4.1 Frequency error

4.1.1 *Definition*

The frequency error of the transmitter is the difference between the measured carrier frequency and its nominal value.

4.1.2 *Method of measurement*

a. The transmitter output in the case of equipment with an antenna terminal, shall be connected to an artificial load (Clause 3.1) and in the case of equipment incorporating integral antenna, shall be placed in the test fixture (Clause 3.2) connected to an artificial load. The transmitter shall be operated in accordance with the manufacturer's instructions to obtain normal output power.

b. The emission shall be monitored by a frequency counter and the carrier frequency shall be measured in the absence of modulation.

c. The measurement shall be made under normal test conditions (Clause 2.3) and repeated under extreme conditions (Clauses 2.4.1 and 2.4.2 applied simultaneously).

4.1.3 *Limits*

The frequency error, under both normal and extreme test conditions, or at any intermediate condition, shall not exceed ± 1.5 kHz. If for determining the transmitter frequency use is made of a synthesizer and/or a phase-locked loop system, the transmitter shall be inhibited when synchronisation is absent.

4.2 Carrier Power

The equipment manufacturer shall provide as a standard accessory an attenuator having a minimum attenuation of 10 dB, or alternatively provide a switch which can be used to reduce the power by a minimum of 10 dB, for use, where necessary, between the transmitter output and the antenna terminals of the equipment, a removable link may be necessary.

4.2.1 *Definition*

For the purpose of this specification: the carrier power shall be the value of the power of an unmodulated carrier at the output terminals of a transmitter. For equipment with an integral antenna, it is the maximum value of effective radiated power of an unmodulated carrier. The rated output power is the maximum value of the transmitter output power declared by the manufacturer, at which all the requirements of this specification are met.

4.2.2 *Method of measurement (Terminal Power)*

a. The transmitter shall be connected to a test load equal to the impedance for which it is designed.

b. With the transmitter operating without modulation in accordance with the manufacturer's instructions, the power delivered to the test load shall be measured.

c. The measurement shall be made under normal test conditions (Clause 2.3) and repeated under extreme test conditions Clauses 2.4.1 and 2.4.2 applied simultaneously.

4.2.3 *Radiated Power*

4.2.3.1 Method of measurement under normal test conditions

a. On a test site fulfilling the requirements of Clause 3.3, the equipment shall be placed on the support in the following position:

 i. equipment with internal antennae shall be arranged with that axis vertical which is closest to vertical in normal use;

 ii. for equipment with rigid external antennae, the antenna shall be vertical;

 iii for equipment with non-rigid external antennae, with the antenna extended vertically upwards by a non-conducting support.

b. The transmitter shall be switched on, without modulation, and the test receiver shall be tuned to the frequency of the signal being measured.

c. The test antenna shall be orientated for vertical polarization and shall be raised or lowered through the specified height range until a maximum signal level is detected on the test receiver*.

d. The transmitter shall then be rotated through 360° until the maximum signal level is received.

e. The transmitter shall be replaced by the substitution antenna, as defined in Clause 3.3 and the test antenna raised or lowered as necessary to ensure that the maximum signal is still received.

f. The input signal to the substitution antenna shall be adjusted in level until an equal or a known related level to that detected from the transmitter is obtained in the test receiver.

g. The carrier power is equal to the power supplied to the substitution antenna, increased by the known relationship if necessary.

h. Steps *a.* to *g.* shall be repeated for any alternative integral antenna supplied by the manufacturer.

j. A check shall be made at other planes of polarization to ensure that the value obtained above is the maximum. If larger values are obtained, this fact shall be recorded in the test report.

4.2.3.2 Method of measurement under extreme test conditions

a. The equipment shall be placed in the test fixture (Clause 3.2) connected to the artificial load (Clause 3.1), with a means of measuring the power delivered to this load.

b. In the absence of modulation, the transmitter shall be operated in accordance with the manufacturer's instructions. The carrier power shall then be measured.

*Note: The maximum may be a lower value than that obtainable at heights outside the specified range.

198

c. The measurement shall be made under normal test conditions (Clause 2.3) and repeated under extreme test conditions (Clauses 2.4.1 and 2.4.2 applied simultaneously).

4.2.4 *Limits*

The carrier power measured under normal test conditions in accordance with Clause 4.2.2 shall not exceed 4 watts. The effective radiated power measured under normal test conditions in accordance with Clause 4.2.3 shall not exceed 2 watts.

The carrier power under extreme conditions shall not exceed by more than 3 dB that measured under normal conditions in accordance with Clause 4.2.2 or 4.2.3 whichever is applicable.

4.3 Frequency deviation

4.3.1 *Definition*

The frequency deviation is the difference between the instantaneous frequency of the modulated radio-frequency signal and the carrier frequency in the absence of modulation. For test purposes, only the maximum value of the frequency deviation available in the transmitter will be measured.

4.3.2 *Maximum permissible deviation*

4.3.2.1 Definition

The maximum permissible frequency deviation is the maximum value of deviation under any conditions of modulation.

4.3.2.2 Method of measurement

a. The equipment, if a fixed station, shall be connected to a test load equal to the impedance for which it was designed and if portable shall be placed in the test fixture (Clause 3.2).

b. The emission shall be monitored by a modulation meter capable of measuring the peak value of both positive and negative frequency deviation including that due to any harmonics and intermodulation products which may be produced in the transmitter.

c. The transmitter shall then be modulated by an audio frequency signal 20 dB above the level necessary to produce normal test modulation (Clause 3.4) and the modulation frequency varied from 300 Hz to 3 kHz.

d. At each test frequency, the peak deviation shall be measured.

4.3.3 *Limit*

At any modulating frequency, the frequency deviation shall not exceed ±2.5 kHz.

4.4 Adjacent channel power

4.4.1 *Definition*

The adjacent channel power is that part of the total power output of a transmitter under defined conditions of modulation, which falls within the bandwidth of a receiver of the type normally used in the system and operating on a channel either 10 kHz above or below the nominal frequency of the transmitter.

4.4.2 Method of measurement

For equipment with radio-frequency output terminals, this measurement shall be carried out at these terminals.

For equipment with integral antennae, this measurement shall be carried out at the output of the test fixture.

a. The equipment or the test fixture shall be connected to the power measuring receiver via a 50 ohm attenuator, set to produce an appropriate level at the receiver input.

b. The transmitter shall be operated at the carrier power measured under normal test conditions in Clauses 4.2.2 or 4.2.3 as applicable.

c. The transmitter shall be modulated at 1250 Hz at a level 20 dB greater than that required to produce 60% of the maximum permissible frequency deviation (Clause 4.3.1).

d. The test receiver shall then be tuned to the nominal frequency of the transmitter and the receiver attenuator adjusted to a value 'p' such that a meter reading of the order of 5 dB above the receiver noise level is obtained.

e. The test receiver shall then be tuned to the nominal frequency of the higher adjacent channel and the receiver attenuator re-adjusted to a value 'q' such that the same meter reading is again obtained.

f. The ratio, in decibels, of the adjacent channel power to the carrier power is the difference between the attenuator settings 'p' and 'q'.

g. The adjacent channel power shall be determined by applying this ratio to the carrier power as determined in Clauses 4.2.2 or 4.2.3 as applicable.

h. The measurement shall be repeated for the lower adjacent channel.

4.4.3 Limits

The adjacent channel power shall not exceed a value of $10\mu.W$

4.4.4 Power measuring receiver specification

The power measuring receiver shall comprise a mixer, a crystal filter, a variable attenuator, an intermediate frequency amplifier, and a r.m.s. meter connected in cascade, using a low noise signal generator as a local oscillator. The bandwidth of the filter shall be as follows (with a tolerance of ±10%):

Bandwidth between 6 dB attenuation points (kHz)	Bandwidth between 70 dB attenuation points (kHz)	Bandwidth between 90 dB attenuation points (kHz)
8.5	17.5	25

The attenuator shall cover a range of at least 80 dB in 1 dB steps. The noise factor of the amplifier shall be not worse than 4 dB. Over the 6 dB bandwidth, the amplitude/frequency characteristics of the amplifier shall not vary by more than 1 dB.

The combined response of the filter and amplifier outside the 90 dB bandwidth shall maintain an attenuation of at least 90 dB. The r.m.s. meter, if not a power meter, shall have a crest factor of at least 10 for the full scale readings. The measuring accuracy of the receiver over an input level range of 100 dB shall be better than 1.5 dB.

4.5   Spurious emissions
4.5.1   *Definition*
Spurious emissions are emissions at frequencies other than those of the carrier and sidebands associated with normal modulation. The level of spurious emissions shall be measured as:

a.   Their power level in a specified load, where the equipment is fitted with output terminals and

b.   Their effective radiated power when radiated by an integral antenna or from the cabinet and chassis of the equipment.

4.5.2   *Method of measurement – power level*
a.   The transmitter output shall be connected to either a spectrum analyser via an attenuator, or an artificial load, with means of monitoring the emission with a spectrum analyser or selective voltmeter.

b.   The transmitter shall be unmodulated and at each spurious emission in the frequency range 100 kHz to 1000 MHz, the level of the emission shall be measured relative to the carrier emission.

c.   The power level of each emission shall be determined by applying the ratio measured to the carrier power level determined in Clause 4.2.3.

4.5.3   *Method of measurement – effective radiated power*
a.   On a test site fulfilling the requirements of Clause 3.2 the transmitter shall be placed at the specified height on the support.

b.   The transmitter shall be unmodulated and its output connected to an artificial load, where the equipment is fitted with output terminals (Clause 3.1)

c.   Radiation of any spurious emissions shall be detected by the test antenna and receiver, over the frequency range 30 to 1000 MHz.

d.   At each frequency at which an emission is detected, the transmitter shall be rotated to obtain maximum response.

e.   The transmitter shall be replaced by a signal generator and dipole antenna and the effective radiated power of the emission determined by a substitution measurement.

f.   The measurements shall be repeated with the test antenna in the orthogonal polarisation plane.

g.   The measurements shall be repeated with the transmitter modulated with normal test modulation (Clause 3.4).

h.   The measurements shall be repeated for any alternative integral antenna which can be supplied with the equipment.

4.5.4   *Limits*
Any spurious emission from the transmitter with and without any ancillary equipment, expressed as a power into a test load or as a radiated power, in either plane of polarisation, shall not exceed 50nW within the following frequency bands:

80 MHz – 85 MHz	174 MHz – 230 MHz
87.5 MHz – 118 MHz	470 MHz – 862 MHz
135 MHz – 136 MHz	

The power of spurious emissions at any other frequency outside the above bands shall not exceed 0.25 μW.

5   Receiver

5.1   Receiver spurious emissions

5.1.1   *Definition*

Spurious emissions from receivers are any emissions present at the input terminals or radiated from an integral antenna or the chassis and case of the receiver.

5.1.2   *Method of measurement for equipment with antenna terminals*

a.   The methods shall be as described in Clauses 4.5.2 and 4.5.3 except that the test sample shall be the receiver.

5.1.3   *Method of measurement for equipment incorporating integral antenna*

a.   The method of measurement shall be as described in Clause 4.5.3 except that the test sample shall be the receiver.

5.1.4   *Limits*

Any spurious emission from a receiver, expressed either as a power into a test load or as a radiated power, shall not exceed 20 nW on any frequency.

6   Accuracy of measurement

The tolerance for the measurement of the following parameters shall be as given below:

6.1.1	DC voltage	±3%
6.1.2	AC mains voltage	±3%
6.1.3	AC mains frequency	±0.5%
6.2.1	Audio-frequency voltage, power etc.	±0.5 dB
6.2.2	Audio frequency	±0.001%
6.2.3	Distortion and noise etc, of audio frequency generators	1%
6.3.1	Radio frequency	±50 Hz
6.3.2	Radio-frequency voltage	±2 dB
6.3.3	Radio-frequency field strength	±3 dB
6.3.4	Radio-frequency carrier power (erp)	±2 dB
6.4.1	Impedance of artificial loads, combining units, cables, plugs, attenuators etc.	±5%
6.4.2	Source impedance of generators and input impedance of measuring receivers	±10%
6.4.3	Attentuation of attenuators	±0.5 dB
6.5.1	Temperature	±1°C
6.5.2	Humidity	±5%

7 Interpretation of this specification

7.1 Application of limits in tests for conformity with this specification. Tests shall be made

7.1.1 either on a sample of appliances of the type using the statistical method of evaluation set out in 7.1.4.

7.1.2 or for simplicity's sake on one item only. The value measured must be at least 2 dB less than the limit value.

7.1.3 Subsequent tests are necessary from time to time on items taken at random from the production especially in the case of 7.1.2. In the case of any dispute which could lead to proceedings under the Wireless Telegraphy Act, such proceedings shall be considered only after tests have been carried out in accordance with 7.1.1.

7.1.4 Statistical assessment of compliance shall be made as follows:

This test shall be performed on a sample of not less than five and not more than 12 items of the type, but if in exceptional circumstances five items are not available, then a sample of three or four shall be used. Compliance is achieved when the following relationship is met:

$$\bar{x} + kS_n \leq L, \text{ where}$$

$\bar{x}$ is the arithmetic mean value of the interference levels on n items in the sample

Sn is the standard deviation of the sample, where

$$S_n^2 = \frac{1}{n-1} \sum (x - \bar{x})^2$$

x is the interference level of an individual item

k is the factor derived from tables of the non-central t-distribution which ensures with 80% confidence that 80% or more of the production is below the limit. Values of k as a function of n are given in the table below.

L is the limit

x, $\bar{x}$, S_n and L are expressed logarithmically

[dB (uV) or dB (pW)]

n	3	4	5	6	7	8	9	10	11	12
k	2.04	1.69	1.52	1.42	1.35	1.30	1.27	1.24	1.21	1.20

7.2 For the purpose of this specification reference to manufacturers, includes importers and assemblers.

1 General

1.1 Scope of specification

This specification covers the minimum performance requirements for angle modulated radio equipments, comprising base station, mobile and hand held transmitters and receivers or receivers only and additionally any accessories, for example attenuators, vehicle adaptors for optional use with the above for use in the Citizens Band Radio service.

For all equipments covered by this specification the nominal separation between adjacent channel carrier frequencies is 25 kHz. (*See note 1*).

1.2 Permitted Effective Radiated Power

The output radio frequency power of the equipment is limited to 8W for equipment which has terminals for connexion of a separate antenna. With the antenna permitted by the conditions of the licence for use with this equipment this provides a maximum effective radiated power of 25W. (*See note 2*).

If an antenna is mounted at a height exceeding 10m the licence will require a reduction in the transmitter power of 10 dB.

To enable the user to accomplish this easily, the equipment manufacturer shall provide as a standard facility on the equipment means by which the transmitter output power may be reduced by a minimum of 10 dB.

For equipment with an integral antenna the effective radiated power is limited to 3W.

1.3 Operating frequencies

The equipment shall provide for transmission and reception only of angle modulated emissions on one or more of the following radio frequency channels:

Channel	1	934.025 MHz	Channel	11	934.525 MHz
,,	2	934.075 ,,	,,	12	934.575 ,,
,,	3	934.125 ,,	,,	13	934.625 ,,
,,	4	934.175 ,,	,,	14	934.675 ,,
,,	5	934.225 ,,	,,	15	934.725 ,,
,,	6	934.275 ,,	,,	16	934.775 ,,
,,	7	934.325 ,,	,,	17	934.825 ,,
,,	8	934.375 ,,	,,	18	934.875 ,,
,,	9	934.425 ,,	,,	19	934.925 ,,
,,	10	934.475 ,,	,,	20	934.975 ,,

Note 1: The initial channel spacing will be 50 kHz.

Note 2: The licence requires that equipments which have provision for the connection of an external antenna, this antenna shall consist of a maximum of four elements none of which may exceed 17cm in length.

Citizens band radio equipment shall not contain facilities for transmission of radio frequencies other than those listed above and those contained in MPT 1320. (*See note 3*).

For the purposes of testing, single channel equipment may be tested on any one of the approved channels. Multi-channel equipment shall be equipped to operate at the centre, and the upper and lower limits of the frequency range over which channel switching is possible.

1.4 Permitted modulation

Only equipment which employs angle modulation and has no facilities for any other form of modulation will meet the requirements of this specification.

1.5 Labelling

The equipment shall be provided with a clear indication of the type number and name of the manufacturer.

1.6 Certification of compliance

Compliance with this specification shall be indicated by a mark stamped or engraved on the front panel of the equipment.

The mark used to indicate compliance shall be as shown in Fig. 1.

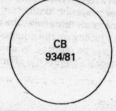

CB
934/81

Letter and figure height not less than 2mm.

Fig. 1.

1.7 Controls

Those controls, which if maladjusted might increase the interfering potentialities of the equipment, shall not be easily accessible.

Note 3: If a synthesiser is used 25 kHz channel spacing may be adopted but only the frequencies corresponding to the numbered channels actually used.

2 Test conditions: Atmospheric conditions and power supplies

2.1 General

Tests shall be made under normal test conditions (Clause 2.3) and also, where stated, under extreme test conditions (Clause 2.4).

2.2 Test power source

During tests, the power supply for the equipment may be replaced by a test power source, capable of producing normal and extreme test voltages as specified in Clauses 2.3.2 and 2.4.2.

The internal impedance of the test power source shall be low enough for its effects on the test results to be negligible.

For the purpose of tests, the supply voltage shall be measured at the input terminals of the equipment.

If the equipment is provided with a permanently connected power cable, the test voltage shall be measured at the point of connection of the power cable to the equipment.

During the tests of the power source voltage shall be maintained within a tolerance of ±3% relative to the voltage at the beginning of each test.

In equipment in which batteries are incorporated, the test power source shall be applied as close to the battery terminals as practicable.

2.3 Normal test conditions

2.3.1 *Normal temperature and humidity*

The normal temperature and humidity conditions for tests shall be any convenient combination of temperature and humidity within the following ranges:

Temperature 15°C to 35°C
Relative humidity 20% to 75%

When it is impracticable to carry out the tests under the conditions stated above, a note to this effect stating the actual temperature and relative humidity during the tests, shall be added to the test report.

2.3.2 *Normal test source voltage*

2.3.2.1 Mains voltage

The normal test source voltage for equipment to be connected to the mains shall be the nominal mains voltage. For the purpose of this specification, the nominal voltage shall be the declared voltage or any of the declared voltages for which the equipment was designed. The frequency of the test power source corresponding to the AC mains shall be between 49 and 51 Hz.

2.3.2.2 Regulated lead-acid battery power sources

When the radio equipment is intended for operation from the usual type of regulated lead-acid battery source, the normal test source voltage shall be 1.1 times the nominal voltage of the battery (6 volts, 12 volts, etc).

2.3.2.3 Other power sources

For operation from other power sources or types of battery, either primary or secondary, the normal test source voltage shall be that declared by the equipment manufacturer.

2.4 Extreme test conditions

2.4.1 *Extreme temperatures*

For tests at extreme temperatures, measurements shall be made in accordance with the procedures specified in Clause 2.5 at an upper value of +45°C and at a lower value of −5°C.

2.4.2 *Extreme test source voltages*

2.4.2.1 Mains voltage

The extreme test source voltages for equipment to be connected to an AC mains source shall be the nominal mains voltage ±10%. The frequency of the test power source shall be between 49 and 51 Hz.

2.4.2.2 Regulated lead-acid battery power sources

When the equipment is intended for operation from the usual type of regulated lead-acid power source, the extreme test voltages shall be 1.3 and 0.9 times the nominal voltage of the battery.

2.4.2.3 Other power sources

The lower extreme test voltage for equipment with power sources using primary batteries shall be as follows:

a. For Leclanche type of battery −0.85 times the nominal voltage
b. For mercury type of battery −0.9 times the nominal voltage
c. For other types of primary battery − end point voltage declared by the equipment manufacturer.

For equipment using other power sources, or capable of being operated from a variety of power sources, the extreme test voltages shall be those declared by the equipment manufacturer and shall be recorded with the test results.

2.5 Procedure for tests at extreme temperatures

2.5.1 *General*

Before making measurements, the equipment shall be placed in a temperature controlled chamber for a period of one hour or so for such period as may be judged necessary for thermal balance to be obtained. The equipment shall be switched off during the temperature stabilisation period. The sequence of tests shall be chosen and the humidity content in the test chamber shall be controlled so that excessive condensation does not occur.

2.5.2 *Test procedure*

For tests at the upper temperature, after thermal balance has been attained (Clause 2.5.1), the equipment shall be switched on for 1 minute in the transmit condition followed by 4 minutes in the receive condition, after which the appropriate tests shall be carried out.

For tests at the lower temperature, after thermal balance has been attained (Clause 2.5.1) the equipment shall be switched on for 1 minute in the receive condition after which the appropriate tests shall be carried out.

208

3 Electrical test conditions

3.1 Transmitter artificial load
Tests on the transmitter shall be carried out using a 50 ohm non-reactive,non-radiating load connected to the antenna terminals. If necessary an impedance matching device may be used for testing.

3.2 Test fixture (integral antenna equipment only)
3.2.1 *General*
A test fixture will be required to permit relative measurements to be made on the sample.*

This test fixture shall preferably provide a 50 ohm radio frequency terminal at the working frequencies of the equipment.

The test fixture shall provide input and output audio coupling and a means of connecting an external power supply.

The following characteristics shall apply to the test fixture.

a. The coupling loss shall be as low as possible, and in any case not greater than 30 dB.

b. The variation of coupling loss with frequency shall not cause errors in measurement exceeding 2 dB.

c. The coupling device shall not incorporate any non-linear elements.

3.2.2 *Stripline arrangement for use as a test fixture*
An example of such a test fixture is given below:

The measuring arrangement consists of an asymmetric parallel plane stripline with width w, mounted at a constant height h above a conducting ground plane by means of plastic supports. For w/h ≥ 4 the width of the ground plane must be at least 3w, for w/h ≤ 3 at least 10w to avoid disturbing effects and to obtain a constant impedance. Under these conditions the impedance depends only on the ratio w/h.

For an impedance of 50 ohms the value of the ratio w/h is 4.95.

In order to provide connection to the stripline the width of either end of the stripline is tapered and its height from the ground plane reduced, to maintain a constant ratio of w/h. Coaxial sockets are connected across the stripline and ground plane at the end of the tapered sections. One end of the stripline shall be matched into a resistive load, and the other end matched into an impedance of 50 ohms.

The VSWR shall be ≤ 1.2 for all frequencies at which measurements are made. Care should be taken that the measuring equipment and any reflecting objects do not disturb the field produced between the stripline and the

*Note: Any connections provided on the equipment in order to facilitate relative measurements, shall not affect the performance of the equipment either in the test fixture or when making measurements involving the use of radiated fields.

ground plane. A small hole can be provided in the centre of the ground plane to enable short connections to the audio frequency circuits.

3.3 Test site and general arrangements for measurements involving the use of radiated fields.

3.3.1 *Test site*

The test site shall be located on a surface or ground which is reasonably level. At one point of the site, a ground plane of at least 5 metres diameter shall be provided. In the middle of this ground plane, a non-conducting support, capable of rotation through 360° in the horizontal plane shall be used to support the test sample at 1.5 metres above the ground plane. The test site shall be large enough to allow the erection of a measuring or transmitting antenna at a distance of not less than half the wavelength corresponding to the lowest frequency to be considered. The distance actually used shall be recorded with the results of the tests carried out on the site.

Sufficient precautions shall be taken to ensure that reflections from extraneous objects adjacent to the site, and ground reflections do not degrade the measurements.

3.3.2 *Test antenna*

The test antenna is used to detect the radiation from both the test sample and the substitution antenna, when the site is used for radiation measurements. This antenna is mounted on a support capable of allowing the antenna to be used in either horizontal or vertical polarization and for the height of its centre above ground to be varied over the range 1–5 metres. Preferably test antenna with pronounced directivity should be used. The size of the test antenna along the measurement axis shall not exceed 20% of the measuring distance.

For radiation measurements, the test antenna is connected to a test receiver, capable of being tuned to any frequency under investigation and to measure accurately the relative levels of signals at its input.

3.3.3 *Substitution antenna*

The substitution antenna shall be a λ/2 dipole, resonant at the frequency under consideration, or a shortened dipole calibrated to the λ/2 dipole. The centre of this antenna shall coincide with the reference point of the test sample it has replaced. This reference point shall be the volume centre of the sample when its antenna is mounted inside the cabinet, or the point where an external antenna is connected.

The distance between the lower extremity of the dipole and the ground shall be at least 0.3m.

The substitution antenna shall be connected to a calibrated signal generator when the site is used for radiation measurements.

The signal generator and the receiver shall be operating at the frequencies under investigation and shall be connected through suitable matching and balancing networks.

3.3.4 *Alternative indoor site*

When the frequency of the signals being measured is greater than 80 MHz, use may be made of an indoor site. If this alternative site is used, this shall be recorded in the test report.*

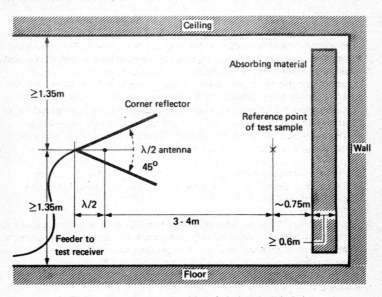

Figure 1   Indoor site arrangement (shown for horizontal polarization)

The measurement site may be a laboratory room having minimum floor dimensions of 6 metres by 7 metres and height of at least 2.7 metres. The room shall be as free as possible from reflecting objects. Potential reflections from the wall behind the equipment under test shall be reduced by interfacing a layer of absorbent material between the test sample and the wall. The corner reflector around the test antenna is used to reduce the effect of reflections from the opposite wall and from the floor and ceiling in the case of horizontally polarised measurements. Similarly, the corner reflector reduced the effects of reflections from the side walls for vertically polarised measurements. For practical reasons, the $\lambda/2$ antenna in Fig 1 may be replaced by an antenna of constant length, allowing it to be used at frequencies corresponding to a wavelength between $\lambda/4$ and $\lambda/2$, as long as the sensitivity is sufficient. In the same way the distance of $\lambda/2$ to the apex

*Note: The requirements for an indoor test site are under review. The details given are an example of such a site which it is considered will give acceptable results.

211

may be varied. The test antenna, test receiver, substitution antenna and calibrated signal generator are used in a similar way to that of the general method.

To ensure that errors are not caused by the propagation path approaching the point at which phase cancellation between direct and remaining reflected signals occurs, the substitution antenna shall be moved through a distance of ±0.1m in the direction of the test antenna as well as in the two directions mutually orthogonal to this first direction. If these changes of distance cause a signal change greater than 2 dB, the test sample should be resisted until less than 2 dB change is obtained.

3.4 Normal test modulation

Where stated, the transmitter shall have normal test modulation as follows: The modulation frequency shall be 1 kHz and the resulting frequency deviation shall be 60% of the maximum permissible frequency deviation (Clause 4.3.1).

4 Transmitter

4.1 Frequency error

The frequency error of the transmitter is the difference between the measured carrier frequency and its nominal value.

4.1.2 *Method of measurement*

a. The transmitter output in the case of equipment with an antenna terminal, shall be connected to an artificial load (Clause 3.1) and in the case of equipment incorporating integral antenna, shall be placed in the test fixture (Clause 3.2) connected to an artificial load. The transmitter shall be operated in accordance with the manufacturer's instructions to obtain normal output power.

b. The emission shall be monitored by a frequency counter and the carrier frequency shall be measured in the absence of modulation.

c. The measurement shall be made under normal test conditions (Clause 2.3) and repeated under extreme test conditions (Clauses 2.4.1 and 2.4.2 applied simultaneously).

4.1.3 *Limits*

The frequency error, under both normal and extreme test conditions, or at any intermediate condition, shall not exceed the values given below:

For all equipments operating in the UHF band the frequency tolerance shall be ±8.0 kHz. If for determining the transmitter frequency use is made of a synthesizer and/or a phase-locked loop system, the transmitter shall be inhibited when synchronisation is absent.

4.2 Carrier Power

4.2.1 *Definition*

For the purpose of this specification: the carrier power shall be the value of the power of an unmodulated carrier at the output terminals of a transmitter. For equipment with an integral antenna, it is the maximum value of effective radiated power of an unmodulated carrier. The rated output power is the maximum value of the transmitter output power declared by the manufacturer, at which all the requirements of this specification are met.

4.2.2 *Method of measurement (Terminal Power)*

a. The transmitter output shall be connected to an artificial load (Clause 3.1) with means of measuring the power delivered to this load.

b. With the transmitter operating without modulation in accordance with the manufacturer's instructions, the power delivered to the test load shall be measured.

c. The measurement shall be made under normal test conditions (Clause 2.3) and repeated under extreme test conditions Clauses 2.4.1 and 2.4.2 applied simultaneously.

st conditions
se 3.3, the equipment
ion:
anged with that axis
use;
ne antenna shall be

ae, with the antenna
ting support.
odulation, and the test
l being measured.
oolarization and shall be
until a maximum signal

60° until the maximum

tion antenna, as defined
d as necessary to ensure

l be adjusted in level until
rom the transmitter is

ed to the substitution
cessary.
ive integral antenna

ation to ensure that the
es are obtained, this fact

est conditions
e (Clause 3.2) connected
easuring the power

hall be operated in
he carrier power shall

test conditions (Clause
lauses 2.4.1 and 2.4.2

ditions in accordance with
radiated power of

ts outside the specified range.

equipment incorporating an integral antenna measured under normal test conditions in accordance with Clause 4.2.3 shall not exceed 3 watts.

The carrier power under extreme conditions shall not exceed by more than 3 dB that measured under normal conditions in accordance with Clause 4.2.2 or 4.2.3 whichever is applicable.

4.3   Frequency deviation
4.3.1   *Definition*
The frequency deviation is the difference between the instantaneous frequency of the modulated radio-frequency signal and the carrier frequency in the absence of modulation. For type approval purposes, only the maximum value of the frequency deviation available in the transmitter will be measured.

4.3.2   *Maximum permissible deviation*
4.3.2.1   Definition
The maximum permissible frequency deviation is the maximum value of deviation under any conditions of modulation.

4.3.2.2   Method of measurement
a.   The equipment, if a fixed station, shall be connected to an artificial load (Clause 3.1) and if portable shall be placed in the test fixture (Clause 3.2) connected to an artificial load (Clause 3.1).

b.   The emissions shall be monitored by a modulation meter capable of measuring the peak value of both positive and negative frequency deviation including that due to any harmonics and intermodulation products which may be produced in the transmitter.

c.   The transmitter shall then be modulated by an audio frequency signal 20 dB above the level necessary to produce normal test modulation (Clause 3.4) and the modulation frequency varied from 300 Hz to 3 kHz.

d.   At each test frequency, the frequency deviation shall be measured.

4.3.3   *Limit*
At any modulating frequency, the frequency deviation shall not exceed ±5.0 kHz.

4.4   Adjacent channel power
4.4.1   *Definition*
The adjacent channel power is that part of the total power output of a transmitter under defined conditions of modulation, which falls within the bandwidth of a receiver of the type normally used in the system and operating on a channel either 25 kHz above or below the nominal frequency of the transmitter.

4.4.2   *Method of measurement**
For equipment with radio frequency output terminals, this measurement shall be carried out at these terminals.

*Note: When using the test fixture for this measurement, it is important to ensure that direct radiation from the transmitter to the power measuring receiver does not affect the result.

For equipment with integral antennae, this measurement shall be carried out at the output of the test fixture.

a. The equipment or the test fixture shall be connected to the power measuring receiver (Clause 4.4.4) via a 50 ohm attenuator, set to produce an appropriate level at the receiver input.

b. The transmitter shall be operated at the carrier power measured under normal test conditions in Clauses 4.2.2 or 4.2.3 as applicable.

c. The transmitter shall be modulated at 1250 Hz at a level 20 dB greater than that required to produce 60% of the maximum permissible frequency deviation (Clause 4.3.1).

d. The test receiver shall then be tuned to the nominal frequency of the transmitter and the receiver attenuator adjusted to a value 'p' such that a meter reading of the order of 5 dB above the receiver noise level is obtained.

e. The test receiver shall then be tuned to the nominal frequency of the higher adjacent channel and the receiver attenuator re-adjusted to a value 'q' such that the same meter reading is again obtained.

f. The ratio, in decibels, of the adjacent channel power to the carrier power is the difference between the attenuator setting 'p' and 'q'.

g. The adjacent channel power shall be determined by applying this ratio to the carrier power as determined in Clauses 4.2.2 or 4.2.3 as applicable.

h. The measurement shall be repeated for the lower adjacent channel.

4.4.3 *Limits*

The adjacent channel power shall not exceed a value of 10 μW.

4.4.4 *Power measuring receiver specification*

The power measuring receiver shall comprise a mixer, a crystal filter, a variable attenuator, an intermediate frequency amplifier, and a r.m.s. meter connected in cascade, using a low noise signal generator as a local oscillator. The bandwidth of the filter shall be as follows (with a tolerance of ±10%):

Bandwidth between 6 dB attentuation	Bandwidth between 70 dB attentuation	Bandwidth between 90 dB attentuation
(kHz)	(kHz)	(kHz)
16	35	50

The attenuator shall cover a range of at least 80 dB in 1 dB steps. The noise factor of the amplifier shall be not worse than 4 dB. Over the 6 dB bandwidth, the amplitude/frequency characteristics of the amplifier shall not vary by more than 1 dB.

The combined response of the filter and amplifier outside the 90 dB bandwidth shall maintain an attenuation of at least 90 dB. The r.m.s. meter, if not a power meter, shall have a crest factor of at least 10 for the full scale readings. The measuring accuracy of the receiver over an input level range of 100 dB shall be better than 1.5 dB.

4.5 Spurious emissions
4.5.1 *Definition*
Spurious emissions are emissions at frequencies other than those of the carrier and sidebands associated with normal modulation. The level of spurious emissions shall be measured as:

a. Their power level in a specified load, where the equipment is fitted with output terminals and

b. Their effective radiated power when radiated by an integral antenna or from the cabinet and chassis of the equipment.

4.5.2 *Method of measurement – power level*
a. The transmitter output shall be connected to either a spectrum analyser via an attenuator, or an artificial load, with means of monitoring the emission with a spectrum analyser or selective voltmeter.

b. The transmitter shall be unmodulated and at each spurious emission in the frequency range 100 kHz to 3800 MHz, the level of the emission shall be measured relative to the carrier emission.

c. The power level of ech emission shall be determined by applying the ratio measured to the carrier power level determined in Clause 4.2.3.

4.5.3 *Method of measurement – effective radiated power*
a. On a test site fulfilling the requirements of Clause 3.3 the transmitter shall be placed at the specified height on the support.

b. The transmitter shall be unmodulated and its output connected to an artificial load, where the equipment is fitted with output terminals (Clause 3.1)

c. Radiation of any spurious emissions shall be detected by the test antenna and receiver, over the frequency range 30 to 3800 MHz.

d. At each frequency at which an emission is detected, the transmitter shall be rotated to obtain maximum response.

e. The transmitter shall be replaced by a signal generator and dipole antenna and the effective radiated power of the emission determined by a substitution measurement.

f. The measurements shall be repeated with the test antenna in the orthogonal polarisation plane.

g. The measurements shall be repeated with the transmitter modulated with normal test modulation (Clause 3.4).

h. The measurements shall be repeated for any alternative integral antenna which can be supplied with the equipment.

4.5.4 *Limit*
Any spurious emission from the transmitter, expressed as a power into a test load or as a radiated power, in either plane of polarisation, shall not exceed 50nW within the following frequency bands:

> 70 MHz – 230 MHz
> 450 MHz – 862 MHz

The power of spurious emissions at any other frequency outside the above bands shall not exceed 0.25 μW.

5 Receiver

5.1 Receiver spurious emissions
5.1.1 *Definition*
Spurious emissions from receivers are any emissions present at the input terminals or radiated from an integral antenna or the chassis and case of the receiver.

5.1.2 *Method of measurement for equipment with antenna terminals*
a. The methods shall be as described in Clause 4.5.2 and 4.5.3 except that the test sample shall be the receiver.

5.1.3 *Method of measurement for equipment incorporating integral antenna*
a. The method of measurement shall be as described in Clause 4.5.3 except that the test sample shall be the receiver.

5.1.4 *Limits*
Any spurious emission from a receiver, expressed either as a power into a test load or as a radiated power, shall not exceed 20nW on any frequency.

6 Accuracy of measurement

The tolerance for the measurement of the following parameters shall be as given below:

6.1.1	DC voltage	±3%
6.1.2	AC mains voltage	±3%
6.1.3	AC mains frequency	±0.5%
6.2.1	Audio-frequency voltage, power etc.	±0.5 dB
6.2.2	Audio frequency	±0.001%
6.2.3	Distortion and noise etc of audio frequency generators	1%
6.3.1	Radio frequency	±50 Hz
6.3.2	Radio-frequency voltage	±2 dB
6.3.3	Radio-frequency field strength	±3 dB
6.3.4	Radio-frequency carrier power (erp)	±2 dB
6.4.1	Impedance of artificial loads, combining units, cables, plugs, attenuators etc.	±5%
6.4.2	Source impedance of generators and input impedance of measuring receivers	±10%
6.4.3	Attentuation of attenuators	±0.5 dB
6.5.1	Temperature	±1°C
6.5.2	Humidity	±5%

7 Interpretation of this specification

7.1 Application of limits in tests for conformity with this specification. Tests shall be made

7.1.1 either on a sample of appliances of the type using the statistical method of evaluation set out in 7.1.4

7.1.2 or for simplicity's sake on one item only. The value measured must be at least 2 dB less than the limit value.

7.1.3 Subsequent tests are necessary from time to time on items taken at random from the production especially in the case of 7.1.2. In the case of any dispute which could lead to proceedings under the Wireless Telegraphy Act, such proceedings shall be considered only after tests have been carried out in accordance with 7.1.1.

7.1.4 Statistical assessment of compliance shall be made as follows: This test shall be performed on a sample of not less than five and not more than 12 items of the type, but if in exceptional circumstances five items are not available, then a sample of three or four shall be used. Compliance is achieved when the following relationship is met:

$$\bar{x} + kS_n \leq L, \text{ where}$$

$\bar{x}$ is the arithmetic mean value of the interference levels on n items in the sample

Sn is the standard deviation of the sample, where

$$S_n^2 = \frac{1}{n-1}\sum (x - \bar{x})^2$$

x is the interference level of an individual item

k is the factor derived from tables of the non-central t-distribution which ensures with 80% confidence that 80% or more of the production is below the limit. Values of k as a function of n are given the table below.

L is the limit

x, $\bar{x}$, S$_n$ and L are expressed logarithmically
[dB (uV) or dB (pW)]

n	3	4	5	6	7	8	9	10	11	12
k	2.04	1.69	1.52	1.42	1.35	1.30	1.27	1.24	1.21	1.20

7.2 For the purpose of this specification reference to manufacturers, includes importers and assemblers.

AUTHOR'S NOTE

Perhaps one of the more interesting things about the new specifications, and of widespread value, is the news that the Home Office have agreed to allow suitably converted pirate AM sets to be used in conjunction with the new FM specification. Several companies have been working on a conversion kit – which consists principally of a new circuit board – and the deal is feasible provided that there is sufficient room inside the casing. The HO will accept these conversions, or private owner conversions, as long as they are taken to the local Customs & Excise Office and meet the requirements of MPT 1320, including fascia panel marking.

Furthermore, and even better news, such converted sets, once a fiver has changed hands in the C&E direction, will be regarded as having been legally imported into the country, which in practice they won't have been. It's a sort of amnesty/free pardon (or nearly free anyway) which is, on the face of it, quite nice.

CB RADIO

A HANDBOOK

BY

RICHARD NICHOLS

The first British CB guide
is presented by acknowledged expert
Richard Nichols, editor of Breaker and
Custom Car.

**Everything you ever wanted
to know about Citizen's Band Radio, including–**

The complete history of CB
and the British and American pirates

A vocabulary of CB
jargon and trucker-talk with 10 codes and
Q codes

The first complete CB map
of Great Britain showing calling channels, clubs
and town handles

Technical topics made easy –
installation, antennae, trouble-shooting for
base and mobile rigs

Read all about the most exciting motoring accessory since the backseat!

THE COMPLETE DICTIONARY OF TRUCKER TALK

CB

LANGUAGE
IN GREAT BRITAIN

CHAS MOORE
(MR BLUE SKY)

You're on the super slab to Brown Ale City and it looks bad. Some linear lungs is jawing on the Japanese set about a Bambi hitting the ton and heading for a bear's den. Your anklebiter's playing up. There's a Harvey Wallbanger letting the hammer down behind you. You could do with some little ponies and maybe a good buddy but most of all you want to know what the hell is going on ...

That's when you need to reach for the keyboard of the lil' ol' modulator and a copy of CB LANGUAGE – the only dictionary of CB talk that can keep your wheels spinning and the beavers grinning!